Phytochemical Arsenal: Understanding Plant Defense Mechanisms Against Nematodes

Edited by

Shivam Jasrotia

Department of Biosciences
Chandigarh University, Mohali
India

&

Ajay Kumar

University Center for Research Development (UCRD)
Chandigarh University, Mohali
India

Phytochemical Arsenal: Understanding Plant Defense Mechanisms Against Nematodes

Editors: Shivam Jasrotia & Ajay Kumar

ISBN (Online): 978-981-5322-67-5

ISBN (Print): 978-981-5322-68-2

ISBN (Paperback): 978-981-5322-69-9

© 2025, Bentham Books imprint.

Published by Bentham Science Publishers Pte. Ltd. Singapore. All Rights Reserved.

First published in 2025.

need for a court order if at any point you breach any terms of this License Agreement. In no event will any delay or failure by Bentham Science Publishers in enforcing your compliance with this License Agreement constitute a waiver of any of its rights.

3. You acknowledge that you have read this License Agreement, and agree to be bound by its terms and conditions. To the extent that any other terms and conditions presented on any website of Bentham Science Publishers conflict with, or are inconsistent with, the terms and conditions set out in this License Agreement, you acknowledge that the terms and conditions set out in this License Agreement shall prevail.

Bentham Science Publishers Pte. Ltd.
No. 9 Raffles Place
Office No. 26-01
Singapore 048619
Singapore
Email: subscriptions@benthamscience.net

BENTHAM SCIENCE

CONTENTS

FOREWORD

In recent years, the field of plant-nematode interactions has gained increasing attention due to the significant economic losses caused by nematode infestations in agricultural crops. As researchers and professionals seek effective and sustainable strategies to manage nematode pests, understanding the intricate defense mechanisms employed by plants becomes paramount. The book "Phytochemical Arsenal: Unveiling Plant Defense Mechanisms Against Nematodes" offers a comprehensive exploration of the fascinating world of plant-nematode interactions and the role of phytochemicals in plant defense.

The unique aspect of this book lies in its focus on the diverse array of phytochemicals produced by plants as their defense arsenal against nematodes. By examining the biosynthesis, mode of action, and ecological significance of these phytochemicals, the authors shed light on the complex interactions between plants and nematodes. The book delves into the mechanisms through which phytochemicals deter, repel, or inhibit nematodes, providing valuable insights into the multifaceted strategies that plants have developed to defend themselves.

With contributions from experts in the fields of plant biology, biochemistry, molecular biology, and agricultural sciences, this book presents a comprehensive and interdisciplinary approach to understanding plant-nematode interactions. It consolidates the latest research findings and scientific advancements, offering a valuable resource for researchers, academicians, and professionals involved in plant pathology, agronomy, crop science, and pest management.

The book goes beyond theoretical discussions by exploring the potential applications of plant-derived phytochemicals in developing sustainable nematode management strategies. It delves into topics such as biopesticides, breeding programs, and biotechnological interventions, providing practical insights into how phytochemicals can be utilized to combat nematode infestations effectively.

As a result of its in-depth exploration of the subject matter, this book serves as an essential guide for researchers and practitioners seeking to enhance their understanding of plant-nematode interactions and develop innovative approaches to nematode management. Its comprehensive coverage, combined with the author's expertise and the integration of case studies and success stories, makes it a valuable resource for both academia and industry.

I commend the authors for their dedication and the wealth of knowledge they have brought together in "Phytochemical Arsenal: Unveiling Plant Defense Mechanisms Against Nematodes." I am confident that this book will be a valuable addition to the scientific literature on plant-nematode interactions and will inspire further research and innovation in the field. It is my pleasure to endorse this book and recommend it to all those interested in advancing our understanding and management of nematode pests in agriculture.

Puja Ohri
Professor & Former Head
Department of Zoology, Guru Nanak Dev University
Amritsar, Punjab 143005, India

PREFACE

In the realm of agriculture, nematodes have long been recognized as formidable adversaries, capable of wreaking havoc on crops and causing significant yield losses worldwide. As the need for sustainable and eco-friendly solutions to combat nematode infestations becomes increasingly urgent, exploring the defense mechanisms employed by plants takes center stage. This book, "Phytochemical Arsenal: Unveiling Plant Defense Mechanisms Against Nematodes," delves into the fascinating world of plant-nematode interactions, with a particular focus on the crucial role of phytochemicals in plant defense.

The journey into the intricate realm of plant-nematode interactions begins with an exploration of the different types of nematodes, their life cycles, and the damage they inflict upon agricultural crops. This foundation sets the stage for understanding the challenges that plant populations face and the importance of unraveling their defense mechanisms against these microscopic parasites.

The book then delves into the diverse and captivating world of phytochemicals, nature's own arsenal of defense compounds produced by plants. Through detailed discussions, readers will gain insights into the various classes of phytochemicals, their biosynthesis pathways, and their roles in defending plants against nematode infestations. From alkaloids and terpenoids to phenolic compounds and glycosides, the book uncovers the intricate chemistry behind these compounds and their impacts on nematode behavior and physiology.

Understanding the mode of action by which phytochemicals combat nematodes is vital in unraveling the intricacies of plant defense mechanisms. This book explores the fascinating interactions between phytochemicals and nematodes, shedding light on how these compounds disrupt nematode feeding, reproduction, behavior, and mobility. Through this exploration, readers will gain a deeper understanding of the remarkable strategies plants have evolved to fend off nematode attacks.

Moreover, the book addresses the influence of environmental factors on the production and efficacy of phytochemicals. It delves into the impact of abiotic and biotic factors on phytochemical synthesis, providing insights into how environmental conditions can be manipulated to enhance plant defense against nematodes.

As researchers and practitioners seek practical applications for their knowledge, the book delves into the exploitation of phytochemicals for nematode management. It examines the potential of phytochemical-based biopesticides, breeding strategies for developing nematode-resistant plants, and biotechnological interventions that harness the power of genetic engineering and beyond. By examining these strategies, the book offers valuable guidance on how to utilize phytochemicals in integrated pest management programs, ultimately contributing to sustainable and effective nematode control.

Throughout the book, case studies and success stories highlight real-world applications and demonstrate the practical implications of phytochemical-based nematode management strategies. These cases provide valuable insights into the challenges faced, lessons learned, and potential pathways for future research and innovation.

In conclusion, "Phytochemical Arsenal: Unveiling Plant Defense Mechanisms Against Nematodes" aims to provide a comprehensive and insightful exploration of the fascinating world of plant-nematode interactions. It consolidates the latest research findings, scientific advancements, and practical applications, making it a valuable resource for researchers, academicians, and professionals in the fields of plant pathology, agronomy, crop science, and pest management.

We hope that this book will serve as an informative and thought-provoking guide, inspiring further research, collaboration, and innovation in the realm of nematode management. We extend our gratitude to all the contributors who have shared their expertise and knowledge, making this book possible. Our sincere hope is that it will contribute to the development of sustainable and effective strategies to protect our crops from the perils of nematodes.

Shivam Jasrotia
Department of Biosciences
Chandigarh University, Mohali
India

&

Ajay Kumar
University Center for Research Development (UCRD)
Chandigarh University, Mohali
India

List of Contributors

Aashaq Hussain Bhat	Department of Bioscience, University Centre for Research and Development, Chandigarh University, Gharuan, Mohali 140413, Punjab, India
Danish Majeed	Department of Botany, University of Kashmir, Hazratbal 190006, Srinagar, Jammu and Kashmir, India
Ewany Jaspher	Experimental Biology Research Group, Faculty of Science, Institute of Biology, University of Neuchatel, Rue Emile Argland 2000, Switzerland
Gagan Preet Kour Bali	Department of Biosciences, UIBT, Chandigarh University, Mohali, Punjab, India
Harsh Gulati	Department of Zoology, School of Bioengineering and Biosciences, Lovely Professional University, Phagwara, Punjab 144411, India
Istkhar Rao	Department of Bioscience and Biotechnology, Banasthali Vidyapith, Banasthali 304022, Rajasthan, India
Ishfaq Majeed Malik	Department of Zoology, University of Kashmir, Hazratbal 190006, Srinagar, Jammu and Kashmir, India
Joginder Singh Rilta	Department of Biosciences, Himachal Pradesh University, Summer Hill, Shimla, India
Kapil Paul	Department of Zoology, Kanya Maha Vidyalaya, Jalandhar, Punjab, India
Kajol Yadav	Department of Biosciences and Biotechnology, Banasthali Vidyapith, Banasthali 304022, Rajasthan, India
Mudasir Ahmad Mir	University Institute of Biotechnology, Chandigarh University, Mohali 140413, Punjab, India
Naveed Nabi	Department of Biosciences, University Centre for Research and Development, Chandigarh University, Gharuan, Mohali 140413, Punjab, Punjab, India
Priyanka Saini	Department of Zoology, Deshbandhu College, University of Delhi, Delhi 110019, India
Puja Ohri	Department of Zoology, Guru Nanak Dev University, Amritsar 143005, India
Prasoon Gupta	Natural Products and Medicinal Chemistry Division, CSIR- Indian Institute of Integrative Medicine, Canal Road, Jammu 180001, India Academy of Scientific and Innovative Research, CSIR-Human Resource Development Centre, Ghaziabad 201002, India
Preety Tomar	Department of Zoology, Eternal University, Baru Sahib, Sirmaur, Himachal Pradesh, India
Raman Tikoria	Department of Zoology, School of Basic Sciences, Central University of Punjab, Bathinda 151401, India
Roohi Sharma	Department of Zoology, Guru Nanak Dev University, Amritsar 143005, India
Ripu Daman Parihar	Department of Zoology, University of Jammu, Jammu & Kashmir 174303, India

Urvashi Dhiman   Natural Products and Medicinal Chemistry Division, CSIR- Indian Institute of Integrative Medicine, Canal Road, Jammu 180001, India
Academy of Scientific and Innovative Research, CSIR-Human Resource Development Centre, Ghaziabad 201002, India

Vaseem Raja   University Centre for Research and Development, Chandigarh University, Gharuan, Mohali 140413, Punjab, India

Overview of Plant-Nematode Interactions and Understanding Plant Defense Mechanisms

Aashaq Hussain Bhat[1,*], Ewany Jaspher[2], Vaseem Raja[3] and Istkhar Rao[4]

[1] *Department of Bioscience, University Centre for Research and Development, Chandigarh University, Gharuan, Mohali 140413, Punjab, India*

[2] *Experimental Biology Research Group, Faculty of Science, Institute of Biology, University of Neuchatel, Rue Emile Argland 2000, Switzerland*

[3] *University Centre for Research and Development, Chandigarh University, Gharuan, Mohali 140413, Punjab, India*

[4] *Department of Bioscience and Biotechnology, Banasthali Vidyapith, Banasthali 304022, Rajasthan, India*

Abstract: Plant-nematode interactions represent a dynamic interplay between parasitic nematodes and their host plants, influencing plant health and agricultural productivity worldwide. This chapter provides a comprehensive overview of the mechanisms underlying plant defense against nematode infestation. It begins with an exploration of nematode parasitism strategies, including sedentary endoparasites and migratory ectoparasites, the discussion delves into the molecular and biochemical mechanisms employed by plants to recognize nematode invasion and mount defense responses. Key topics include the role of plant hormones such as jasmonic acid and salicylic acid in signaling pathways, the activation of defense-related genes, and the induction of physical barriers to nematode penetration. Furthermore, recent advances in understanding plant-nematode interactions, such as the discovery of nematode effectors and their manipulation of plant immunity, are highlighted. Additionally, the chapter examines the potential application of biotechnological approaches, such as breeding for nematode resistance and the use of biocontrol agents, in managing nematode infestations sustainably. By elucidating the intricate mechanisms of plant defense against nematodes, this chapter aims to contribute to the development of effective strategies for enhancing crop resilience and ensuring global food security. By synthesizing current knowledge and research findings, this chapter contributes to a comprehensive understanding of plant-nematode interactions and provides insights into novel avenues for enhancing plant resistance to nematode pests. Ultimately, elucidating the intricacies of plant defense mechanisms against nematodes holds promise for sustainable agriculture practices and the development of resilient crop varieties in the face of evolving pest pressures.

* **Corresponding author Aashaq Hussain Bhat:** Department of Bioscience, University Centre for Research and Development, Chandigarh University, Gharuan, Mohali 140413, Punjab, India; E-mail: aashiqhussainbhat10@gmail.com

Keywords: Plant immunity, Nematode effectors, Plant-nematode interactions, Plant defense mechanisms, Molecular signaling pathways.

INTRODUCTION TO PLANT-NEMATODE INTERACTIONS

Plant-parasitic nematodes (PPNs) are small microscopic creatures that pose significant threats to agriculture and horticulture by feeding on the roots of plants, leading to decreased crop productivity, hindered growth, and occasionally plant demise. They represent a great problem in global crop production as they bring about substantial economic losses every year [1 - 4]. Among these, inactive endoparasitic nematodes under the family Heteroderidae, such as root-knot nematodes (RKNs) (*Meloidogyne* spp.) and cyst-forming nematodes (CNs) (*Heterodera* and *Globodera* spp.), are particularly troublesome [3, 5].

CNs typically have a limited range of hosts, while RKNs are known for their ability to parasitize a wide variety of flora on this planet [6]. Among CNs, the most harmful species are primarily those belonging to the *Heterodera* and *Globodera* genera [7]. Four prevalent and adaptable RKN species are *Meloidogyne incognita*, *M. javanica*, *M. arenaria* (apomictic), and *M. hapla* (automictic), which cause substantial agricultural damage. Both CNs and RKNs remain within the plant roots for a major part of their life cycle, typically completing their reproductive cycle within 2-3 fortnights [8]. They possess specialized structures allowing efficient penetration of plant tissues, modification of root cells, and extraction of nutrients for their growth and reproduction. These nematodes employ complex mechanisms to parasitize host plants, leading to the formation of specific nourishing sites—giant cells for RKNs and syncytia for CNs—within plant roots. Giant cells arise from the transformation of approximately six vascular root cells, undergoing repeated division of their nuclei not followed by cell division, resulting in enlarged cells with multiple nuclei that can be more than 300 times larger than typical cells. Surrounded by dividing cells, these enlarged cells lead to the creation of characteristic galls [9 - 11]. The formation of a syncytium begins when the root cell wall partially dissolves and the protoplasts of the first infected vascular cell merge with adjacent cells. This process can lead to the formation of syncytia that include more than 200 cells. Syncytia, like giant cells, share several traits such as an expanded endoplasmic reticulum, broken down vacuoles, rearranged cytoskeletons, reinforced cell walls with localized protrusions, extensive mitochondrial networks, and nuclei that have undergone endoreduplication [10, 12]. These specialized nematode feeding sites (NFS) are crucial as they supply nutrients to the nematodes during the non-mobile stages of their life cycle.

Female RKNs deposit their eggs encased in a gel-like substance on plant roots, whereas CNs embed their eggs within the hardened body of the dead female, forming a cyst. These parasites induce several changes in the morphology, biochemistry, and molecular structure of plant root cells to create nourishing sites. This alteration is primarily driven by substances the nematodes emit through their stylets, which trigger a series of signaling events within the host cells [10, 12]. The interaction between the nematodes and the plants begins when the second-stage juvenile nematodes (J2) move towards the roots, attracted by substances the roots emit [13]. In the absence of defensive reactions from the plant, such as the production of reactive oxygen species (ROS) and the deposition of callose, the nematodes penetrate the root cells and target the vascular system to establish their NFS [13]. These sites then serve as the main nutritional sources for the nematodes as they continue their development. Nematodes secrete specific molecules, referred to as "effectors," that help them invade the host roots, evade plant defenses, and modify the root cells to create specialized NFS [13]. These effectors are primarily produced in the nematodes' three esophageal salivary glands and delivered into plant cells *via* the stylet, which functions like a syringe. The activity of these glands changes over time; initially, the two subventral glands (SvG) are primarily active in aiding J2 penetration and movement within the root, but as the nematode matures, secretion mainly occurs from these glands and particularly the dorsal gland (DG) [13, 14]. Additionally, effectors can also come from other secretory structures like the chemosensory amphids or be directly secreted through the nematode's cuticle. Research has mainly focused on the protein-based effectors [15 - 19], though there are indications that other molecules such as phytohormones also significantly impact these interactions [20].

Nematode secretions are essential for initiating the formation of NFSs, containing a variety of components such as CWDEs, Avr proteins, and transcription factors [21, 14]. Through these effector proteins, PPNs interact with host genes and proteins, facilitating successful parasitism [21 - 23]. These secretions induce significant changes in host gene regulation *via* the stylet, prompting the conversion of parasitized root cells into specialized NFS [24]. Moreover, nematode effectors can manipulate plant metabolic and developmental pathways to induce NFS formation and suppress host defense mechanisms [21, 23]. The initiation of NFSs begins with J2s hatching from eggs and migrating towards roots. They initiate infection by penetrating the epidermal cells of the root, using their stylet to create openings in the cell wall (Fig. **1**). These nematodes then navigate through the root's cortical cells toward the vascular cylinder. The choice of ISC varies by species; for instance, *Globodera* species often target cells in the inner cortex or endodermis [25, 26], whereas *H. schachtii* commonly selects cells in the cambium or procambium of the vascular cylinder [27]. Using their stylet, J2s breach the ISC's cell wall and inject esophageal gland secretions into the cytoplasm of the plant cell. These secretions include effector proteins that alter the plant's developmental pathways to support the formation of a syncytium [27]. These secretions, originating from the pharyngeal glands, contain effector proteins that modify plant morphogenetic pathways, facilitating syncytium development [16]. This process leads to the merging of adjacent cells into the ISC, dissolving their cell walls locally, resulting in a large, multinucleate syncytium formed from what were originally separate root cells [28, 29].

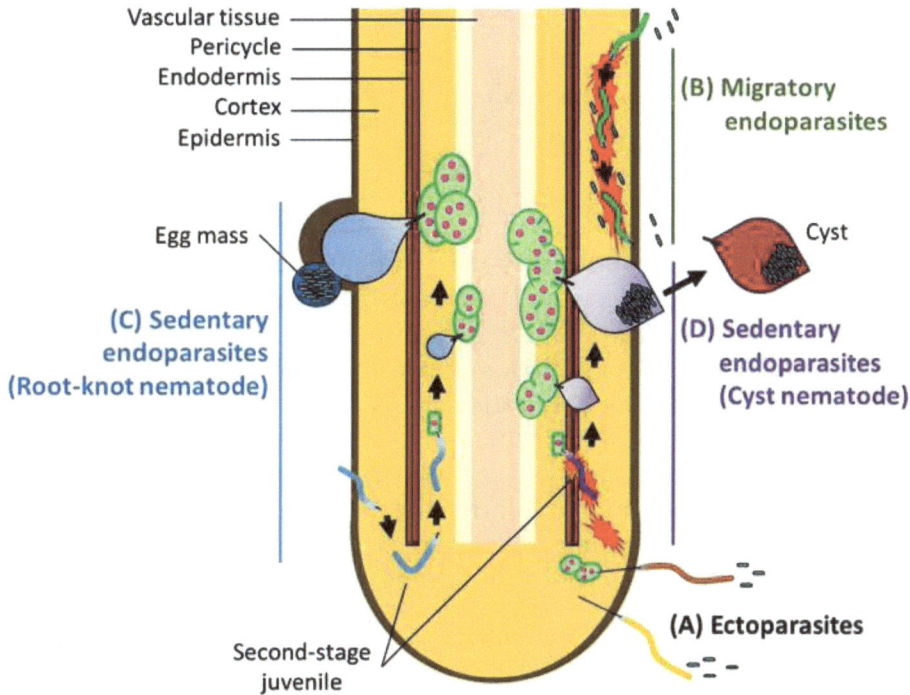

Fig. (1). Infection strategies of PPNs. (**A**) Ectoparasites extract nutrients externally from plant roots. (**B**) Migratory endoparasites traverse and damage root tissues while feeding. (**C**) RKN second-stage juveniles infiltrate near the root-tip, form giant cells in the vascular cylinder, and then feed sedentarily. Adult RKNs produce egg masses on/under the root surface. (**D**) CN second-stage juveniles penetrate root cells, create syncytia in vascular tissues for feeding, and become sedentary. Adult CNs form a cyst post-mortem by retaining eggs internally [30].

Understanding plant-nematode interactions is crucial for several reasons. Firstly, nematodes are microscopic organisms that can cause devastating damage to crops, leading to significant economic losses in agriculture globally. By comprehending the dynamics of these interactions, researchers can develop effective strategies for nematode management, such as breeding resistant plant varieties or implementing sustainable pest control methods. Secondly, these interactions are integral to ecological systems, influencing soil health, nutrient cycling, and overall ecosystem stability. Understanding the intricate relationships between plants and nematodes sheds light on the broader ecological processes and aids in the development of environmentally friendly agricultural practices. Additionally, insights gained from studying plant-nematode interactions can contribute to advancements in biotechnology, including the development of novel biocontrol agents or biologically based approaches for pest management, thus paving the way for more sustainable and resilient agricultural systems. Overall, unravelling the complexities of plant-nematode interactions holds immense importance for safeguarding food security, preserving ecosystems, and advancing agricultural innovation.

UNDERSTANDING THE SEDENTARY ENDOPARASITIC NEMATODE LIFE CYCLE

The life cycle of non-motile endoparasitic nematodes commences when J2, hatching from eggs (J1), becomes active in the soil. These J2 juveniles employ sensory organs known as amphids to detect chemical cues within root exudates, aiding in the identification of a suitable host [31]. Their primary objective is to find and infiltrate host tissue before their energy, particularly stored fats, runs out. To penetrate the host, J2 nematodes apply physical force with a needle-like structure called a stylet to break through the plant cell walls. During this process, they also release a variety of enzymes that further break down the cell walls, facilitating their entry and movement within the host. Migration can occur intracellularly (for CN) or intercellularly (for RKN), as they progress towards the vascular bundle. CNs, emerging as J2 from their eggs, embark on a quest for a suitable host, penetrating its root system and navigating within the root cortex until discovering a suitable feeding cell proximate to the vascular bundle. CNs manipulate host cell processes by releasing certain chemicals, leading to the partial breakdown of the plant cell walls and the merging of nearby cells. This process results in the creation of a syncytium, a large, nutrient-rich structure that the nematode uses for sustenance [32]. RKNs, particularly in their J2 stage, penetrate the roots near the base and navigate between cells toward the root's growing tip [33]. Once they approach the tip, RKNs alter their path, heading back to the developing vascular system. Here, they initiate the development of NFS by injecting their secretions into several nearby pre-vascular cells using their stylet [34]. These targeted cells undergo repeated cycles of nuclear division without cell division, leading to the emergence of enlarged, multinucleate cells known as giant cells. These giant cells, which can grow up to 100 times their original size, serve as a nutrient depot for the nematode, typically encompassing five to seven such cells per nematode.

Upon reaching the vascular bundle, nematodes target specific xylem parenchymal cells, prompting them to create specialized NFS. After settling in, the nematode stops moving and goes through a series of molts, progressing from J2 to J3 and J4, before ultimately maturing into an adult (Fig. **2**). Mature males exhibit elongated, mobile, worm-like forms and exit the plant, while mature females display a rounded, pear-shaped morphology, equipped with a small feeding structure facilitating nutrient intake. In case of CNs, females engage in sexual reproduction, yielding a substantial number of eggs [35]. Upon the female's demise, the body wall forms a protective cyst around the eggs, which are then extruded from the roots, giving rise to the term "cyst nematodes." In contrast, RKN females reproduce *via* parthenogenesis, generating numerous eggs ensconced in a gel-like substance referred to as egg masses [36]. These egg masses can be observed on the root knots, as they are discharged from the roots, while the female stays embedded within. This leads to the commencement of a new infection cycle when the eggs hatch into J2 juveniles [37].

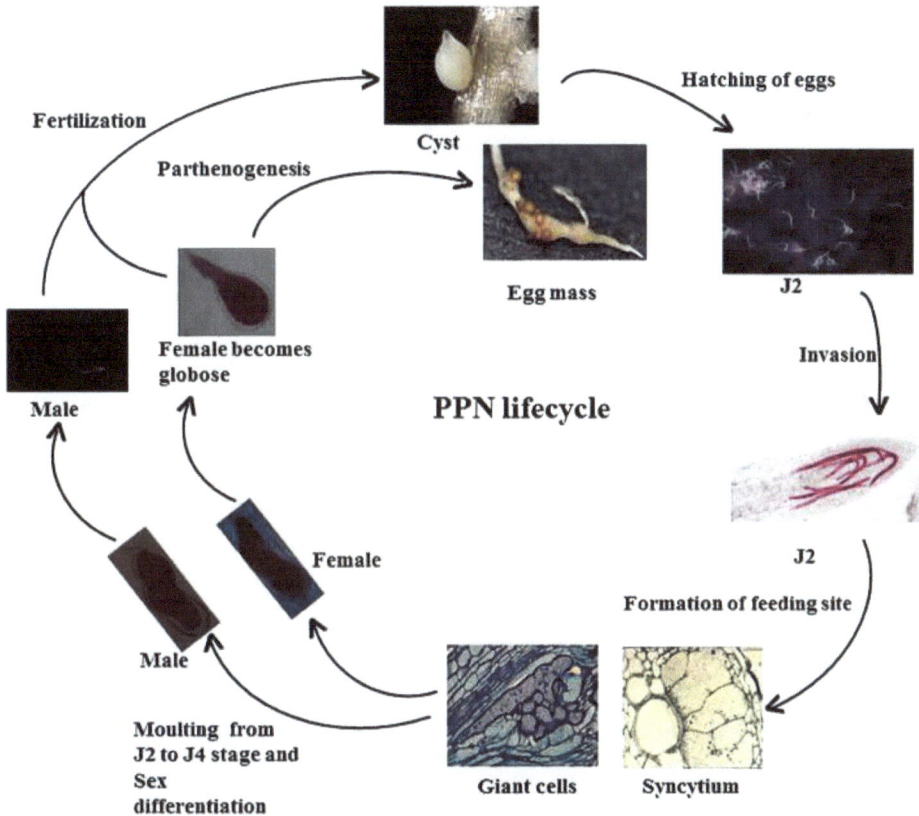

Fig. (2). The developmental stages of stationary plant parasitic nematodes, specifically root-knot nematodes (RKNs) and cyst nematodes. Root-knot nematodes induce the formation of giant cells within the roots of the host plants and reproduce asexually, resulting in the creation of clusters of eggs. In contrast, cyst nematodes generate specialized structures called syncytia within the host plant roots and reproduce sexually, leading to the development of protective cysts containing their eggs [38].

NFS represent highly active, sizable, multi-nucleated entities characterized by dense granular cytoplasm, serving as the primary nutritive reservoir for embryonic nematodes. They are denoted as syncytia in CNs [39] and as colossal cells in RKNs [34]. Colossal cells develop from the activation of cell division (karyokinesis) without subsequent cell separation (cytokinesis) in a small number of vascular parenchymal cells, typically 5–7. This process induces hyperplasia and hypertrophy in the surrounding cortical tissues, resulting in the formation of characteristic galls known as root-knots [12]. Conversely, giant cells also form from the stimulation of karyokinesis without cytokinesis in 5–7 vascular parenchymal cells (Fig. **3**), leading to hyperplasia and hypertrophy in the adjacent cortical tissues and the subsequent development of root-knot galls [12]. Nematodes have evolved mechanisms for effective communication with their host plants, adapting these interactions to meet their own growth and reproductive needs. While earlier studies have shed light on various aspects of nematode behavior, morphology, physiology, genetics, and life cycles, recent efforts have been predominantly directed towards understanding the molecular underpinnings of plant-nematode interactions.

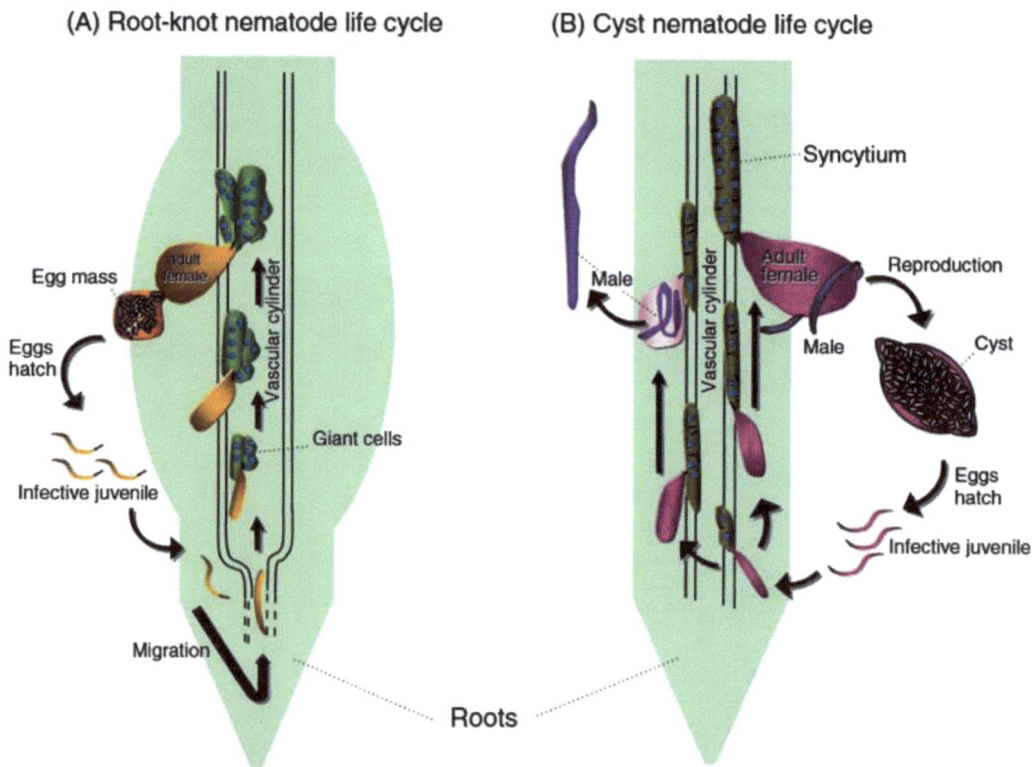

Fig. (3). The life cycle of (**A**) RKN and (**B**) CN [37]

ENZYMATIC SOFTENING BY NEMATODES AND PLANT DEFENSE SUPPRESSION

PPNs employ an array of tactics to circumvent plant defenses and enhance their feeding. One such strategy entails the exudation of a concoction of enzymes that dismantle the plant cell walls' structural integrity. This enzymatic amalgamation renders plant tissues more pliable, thereby facilitating the infiltration of roots or other plant structures for nutrient extraction, ultimately culminating in detrimental effects or symptoms of disease in the host plant. Nematode stylet secretions encompass a diverse ensemble of enzymes, including pectate lyases, expansins, β-1,4-endoglucanases, polygalacturonases and xylanases [21, 40, 41]. These enzymes function by catalyzing the hydrolysis of cellulose, hemicellulose, and pectin constituents, consequently dismantling and altering the cell wall matrix. The development of fluorescent protein tagging has significantly advanced our understanding of where these substances localize within the host tissue. The discovery of β-1,4-endoglucanase as an initial nematode effector in plant roots, first found in *Heterodera glycines*, marked a critical milestone in plant-pathogen interactions [41]. This event was particularly notable as it represented the first time a cellulase gene was identified in any animal. The structural similarities between nematode cellulases and those of bacteria and fungi have led researchers to propose that nematodes might have acquired these enzymes *via* horizontal gene transfer [42, 43]. Furthermore, the cellulose binding protein (CBP) from *Heterodera schachtii* has been shown to interact with pectin methyl esterase 3 in *Arabidopsis*, decreasing the levels of methyl-esterified pectin in the plant cell walls. This interaction weakens the cell walls and increases plant vulnerability to nematode attack [44]. Expansins play a different role by loosening non-covalent links within cell wall fibrils, which helps relax the structure of these walls [45]. Additionally, these nematodes produce proteases that not only aid in cell wall loosening to ease their movement but also break down plant defense proteins and digest proteins within the giant cells formed by the host [45].

Upon breaching root tissues, nematodes encounter an array of plant defense strategies, such as generating ROS, strengthening cell walls, inducing lignification, and producing secondary metabolites. In response, plants deploy chemical defenses, such as the synthesis of toxic compounds or the release of signaling molecules activating immune responses [46]. Nematodes have evolved various strategies to counteract these defenses, thereby dampening host defense responses to ensure successful parasitism. One such strategy involves the action of chorismate mutase (CM) enzymes, primarily located in esophageal glands, which actively contribute to suppressing plant defense mechanisms [21]. These nematode CM enzymes target chorismate substrates, inhibiting the synthesis of crucial components of plant defense pathways, including phytoalexins, auxin, flavonoids, and salicylic acid [47]. Additionally, FAR-1, belonging to the fatty acid and retinol-binding protein family, engages with precursors in the jasmonic acid pathway, thereby inhibiting plant defense mechanisms [48]. Proteins called peroxidases, which help detoxify ROS, have been found on the surface of the hypodermis in potato cyst nematodes (PCNs) [49]. Additionally, genes encoding glutathione-S-transferase (GST), identified in the endophytic J3 nematodes, are thought to inhibit plant defense mechanisms at both the plasma membrane and cell wall levels during various phases of parasitism [50]. Overall, the enzymatic softening orchestrated by nematodes and their adept suppression of plant defenses represent intricate strategies employed to surmount the barriers presented by host plants and exploit them as a nutrient source. Grasping these dynamics is essential for developing successful approaches to manage nematode populations and protect agricultural crops.

PLANT DEFENSE MECHANISMS AGAINST NEMATODES

Plants have developed an array of defense barriers against PPNs, which pose significant threats to root health. These mechanisms primarily involve physical barriers and chemical defenses. Physical barriers are essential as the initial form of protection, playing a key role in preventing PPN infections. For example, plants may strengthen their root tissues with thickened cell walls, making it difficult for PPN stylets, specialized mouthparts for feeding, to penetrate. Additionally, plants can create physical barriers such as corky layers or lignified tissues around the roots, hindering nematode penetration and nutrient access. In resistant plants, PPN infection triggers the accumulation of lignin, strengthening the cell wall structure and improving its effectiveness as a physical barrier [51 - 53]. Studies by Wuyts *et al.* (2006) have revealed that heightened levels of syringyl lignin diminish RKN reproduction rates, underscoring the role of lignin as a potent antagonist against PPN infection in roots [54].

Additionally, plants employ supplementary defensive mechanisms to discourage infestation by PPN. Immunomodulatory substances like sclareol, β-aminobutyric acid (BABA), and thiamine, have shown efficacy in thwarting PPN infection. For instance, BABA application interferes with RKN invasion, postpones the formation of giant cells, and hinders RKN growth [55]. Thiamine application promotes the buildup of lignin in roots, increases the activity of phenylalanine ammonia-lyase—a key enzyme in the phenylpropanoid pathway—and reduces PPN invasion and growth [56]. Furthermore, sclareol application has been shown to enhance lignin deposition and inhibit RKN penetration [56]. Notably, the significance of lignin accumulation in sclareol-mediated immunity is underscored by research showing that an *Arabidopsis* mutant deficient in lignin accumulation fails to induce RKN penetration suppression by sclareol [57].

The reinforcement of plant cell walls through the accumulation of callose and suberin boosts their defense against PPNs. Notably, in the robust grass *Aegilops variabilis*, early stages of infection by the RKN *Meloidogyne naasi* trigger the formation of callose, while later stages promote the development of suberin [52]. The importance of suberin in blocking PPN penetration is emphasized by studies showing that *Arabidopsis* mutants without Casparian strips, which rely on suberin, are more vulnerable to RKNs [33]. Additionally, the overexpression of the transcription factor RAP2.6, which increases callose deposition around syncytia, correlates with increased resistance to CNs [58]. The significance of suberin in thwarting PPN infiltration is demonstrated by studies revealing that *Arabidopsis* mutants deficient in Casparian strips, a suberin-derived defense, show heightened vulnerability to RKNs [33]. Hence, lignin, callose, and suberin, acting as physical barriers, play pivotal roles in plant defense against PPNs [59, 60].

Plants initiate the production of secondary metabolites like chlorogenic acid in response to attacks by PPNs. This phenolic compound is synthesized by a variety of plants including those in the Solanaceous family [61, 62], as well as carrots [63], and rice [64] suggesting a widespread defense mechanism against PPN invasions. Although chlorogenic acid has been linked to PPN resistance, its effectiveness against *Meloidogyne incognita* is somewhat limited [65]. Conversely, it shows a better performance in controlling *Nacobbus aberrans*, another type of root-knot nematode [66]. This discrepancy might be explained by the possible nematicidal properties of the breakdown products of chlorogenic acid. Despite their potential, these derivatives, such as caffeic acid which degrades into orthoquinone, might be either too unstable or harmful to the plants themselves. Orthoquinone, in particular, is known to have detrimental effects on PPNs [66]. Further studies are required to clarify the exact contributions of these metabolites in combating PPNs.

In banana cultivars of the *Musa* sp. that are resistant to damage, the phenolic compound known as phenylphenalenone anigorufone has been found to accumulate at locations affected by infections caused by the burrowing nematode, *Radopholus similis* [53, 67]. This compound manifests potent nematicidal attributes by engendering substantial lipid-anigorufone amalgams within the anatomies of *R. similis*. Additionally, it functions as a phytoalexin with antifungal efficacy, effectively combating infections from the fungus *Fusarium oxysporum* [68]. Moreover, anigorufone is toxic to the human protozoan parasite *Leishmania*, as it inhibits succinate dehydrogenase in the mitochondrial respiratory complex II [69]. However, a deeper understanding of how anigorufone specifically impacts PPNs and the relationship this has with the formation of lipid-anigorufone complexes is essential.

Flavonoids, an assorted array of secondary botanical metabolites, assume pivotal roles in combatting PPN. They serve multifarious functions, including acting as nematicidal agents, impeding locomotion, repulsing, and impeding the hatching of eggs [70]. Specifically, types of flavonoids like flavonols (for example, quercetin, myricetin and kaempferol), isoflavonoids, and pterocarpans (such as glyceollin and medicarpin) are valued for their abilities to combat nematodes Kaempferol, for instance, has been observed to prevent the hatching of *R. similis* eggs [71]. Additionally, medicarpin has shown a dose-dependent reduction in the movement of *Pratylenchus penetrans* [72]. Compounds like rutin, quercetin, patulitrin, and patuletin demonstrate lethal effects on the juvenile stages of *Heterodera zeae* [73]. Some flavonoids are produced in higher quantities in plant varieties resistant to *M. incognita* following infection. For example, glyceollins, specific to soybeans and classified as prenylated pterocarpan phytoalexins, are synthesized in response to an infection [74]. Glyceollin has been found to restrict the movement of *M. incognita* [74, 75], particularly accumulating around the nematode's head region in the roots of resistant soybean varieties, indicating a targeted response to the presence of the pathogen.

In addition to phenolic compounds, various plants known for their antagonistic effects against nematodes also synthesize alternative nematicidal substances. For instance, marigold roots secrete α-terthienyl, a compound documented by Faizi *et al.* (2011) and Sánchez-Sánchez & Morquecho-Contreras (2017), which induces oxidative stress in nematodes [73, 76]. This chemical induces oxidative stress in nematodes and exhibits substantial nematicidal efficacy [77]. Similarly, *Asparagus* plants generate asparagusic acid, which, according to Takasugi *et al.* (1975) effectively prevents the eggs of two major cyst nematodes, *Heterodera glycines,* and *G. rostochiensis*, from hatching [78].

Plants in the Brassicaceae family are recognized for producing strong antimicrobial compounds such as isothiocyanates and indole glucosinolates, which are effective against PPNs. Studies have shown that isothiocyanates can prevent the eggs of RKNs and CNs [79], and are lethal to several nematode species, including RKNs and *Tylenchulus semipenetrans* [80]. In *Arabidopsis*, key enzymes involved in the production of the indole alkaloid glucosinolate phytoalexin known as camalexin include the cytochrome P450-dependent monooxygenases (CYP79B2, PAD3, and CYP79B3) [81 - 84]. Research indicates that *Arabidopsis* mutants that do not accumulate enough indolic glucosinolates are more prone to attacks by CNs [85], and those that cannot produce camalexin are more susceptible to RKNs [86]. These observations suggest that specific indole glucosinolates, such as camalexin, might have protective effects against PPNs, though the direct toxic impact of these substances on PPNs has yet to be fully established.

PPNs can be effectively controlled by interfering with their chemotactic responses towards plant roots. Ethylene (ET), a chemical produced by plants in response to damage or pathogen attack, reduces the attraction of PPNs to roots [87 - 89]. Research shows that *Arabidopsis* plants genetically modified to produce excess ET are less attractive to PPNs. Conversely, plants that either inhibit ET production or are resistant to ET show increased susceptibility to PPN attraction [90, 91]. These studies suggest that ET production triggered by PPN attacks could serve as a defensive mechanism to reduce further invasions by decreasing root attractiveness to PPNs.

Molecular Signaling Pathways

Plants have evolved sophisticated mechanisms to defend themselves against various pathogens, including nematodes. Key to these defenses are molecular signaling pathways that manage the response to nematode invasions. Notably, the pathways involving jasmonic acid (JA), salicylic acid (SA), and ethylene (ET) are critical in these processes [92 - 95].

JA is a critical plant hormone that orchestrates defense mechanisms against various environmental threats such as herbivory by insects and attacks by necrotrophic pathogens, including certain nematodes [93, 96 - 98]. During a nematode attack, the synthesis of JA is initiated *via* the octadecanoid pathway. In plants like *Arabidopsis*, this hormone combines with isoleucine, facilitated by the enzyme jasmonoyl isoleucine conjugate synthase1 (JAR1), to form a complex with the protein coronatine insensitive 1 (COI1). This interaction leads to the breakdown of jasmonate ZIM domain (JAZ) repressor proteins, thereby activating genes responsive to JA [99]. Activation of these genes includes those coding for the negative regulators such as JAZ proteins [93]. Two critical pathways have been identified: MYC2 which upregulates genes like vegetative storage protein 2 (VSP2) and lipoxygenase 2 (LOX2), both of which respond to JA and are also triggered by mechanical damage. The release of JA and its related compounds during nematode invasion sets off a series of defensive responses, which include the generation of harmful metabolites and protective proteins. Additionally, the JA signaling pathway enhances the plant's resistance to nematodes by boosting cell wall fortification and the synthesis of other defense hormones [100, 101].

SA plays a crucial role in the plant immune system, acting as a key signal transmitter when pathogens are detected by various immune receptors. This signaling initiates defense responses particularly effective against biotrophic and hemi-biotrophic pathogens, as well as certain nematodes. This function of SA in plant defense is well documented in various studies [96 - 98, 102 - 104]. In particular, when nematodes penetrate plant roots, SA production is stimulated, which activates defense mechanisms, including the release of pathogenesis-related (PR) proteins with antimicrobial effects that restrict nematode growth [105, 106]. These defense actions in plants are controlled by the SA pathway, which converts chorismate to SA through both the isochorismate and phenylalanine ammonium lyase pathways [107]. Increased levels of SA lead to the nuclear translocation of the pathogenesis-related gene 1 (NPR1), which in turn promotes the expression of defense proteins like PR proteins [108]. Moreover, when plants face attacks from multiple herbivores, the interplay among JA, SA, and ET signaling pathways regulates the defensive responses. This interaction often results in mutual suppression between the JA and SA pathways, prioritizing one pathway's defensive strategy over the other [109]. Furthermore, the SA pathway influences the generation of ROS, adding to the oxidative burden that nematodes experience within plant tissues.

The ET pathway signaling pathway is crucial for plant defenses against various nematode species, although its effectiveness varies depending on the specific nematode and plant involved. ET, a volatile phytohormone, is produced in response to multiple stress factors, including attacks by nematodes. This hormone initiates a series of signaling events that lead to changes in plant physiological and structural aspects, such as the modification of root structures and the triggering of genes associated with defense [87 - 89]. For instance, *Arabidopsis* mutants that overproduce ET are less attractive to PPNs, while blocking ET synthesis or using mutants insensitive to ET increases susceptibility [90, 91]. This suggests that the ET surge following nematode infection may serve as a deterrent against further PPN invasions by making the plants less appealing. Additionally, ET can interact with other signaling molecules such as JA and SA to optimize the plant's defense strategy against nematode incursion. Within the ET signaling framework, the ET response factor (ERF) activates ERF1 and ORA59, both of which are transcription factors responsive to JA and ET and play roles in regulating genes like plant defensin 1.2 (PDR1.2) [92]. Meanwhile, MYC2 plays a significant role in herbivore defense, whereas ERF is more crucial in defending against necrotrophic pathogens [110].

Overall, these signaling pathways are vital elements of a plant's protective mechanisms against nematodes. They orchestrate elaborated biochemical and genetic reactions that enable plants to effectively counter nematode attacks, thereby preserving their structural integrity. Gaining insights into how these pathways work together is critical for devising methods to boost plant immunity to nematodes, thereby enhancing agricultural yield.

Role of Plant Genetics in Nematode Resistance

Plant genetics plays a pivotal role in conferring resistance to nematode infestations, providing farmers and breeders with potent tools to combat these pests effectively. Developing nematode-resistant plant varieties is imperative for ensuring sustainable agriculture and enhancing food security. Plant genetics contributes significantly to nematode resistance by leveraging naturally occurring resistant traits within plant species. Some plants have developed protective strategies to combat nematode infestations. These strategies include preventing nematodes from entering the plant, hindering their ability to reproduce, or minimizing the impact of their feeding activities. By identifying the genes responsible for these resistant traits, breeders can integrate them into commercial crop varieties using traditional breeding methods or modern genetic engineering techniques.

Identification of Resistance Genes

Research on resistance genes is key to unraveling how plants defend themselves against nematodes. This field focuses on the detection, description, and application of genes that provide nematode resistance. Recent developments in genomic technology, transcriptomic analysis, and bioinformatics have significantly enhanced our ability to identify and examine these crucial genes in plants.

Plants employ basal defense mechanisms, termed pattern-triggered immunity (PTI), which are activated through the engagement of pattern recognition receptors (PRRs) with pathogen-associated molecular patterns (PAMPs) to block the spread of pathogens and limit infections [111, 112]. Despite this defense, some virulent pathogens manage to bypass PTI by discharging effector molecules that induce effector-triggered susceptibility (ETS) [113]. To counteract this, plants have developed resistance genes that detect these specific effectors, referred to as avirulence (Avr) proteins, activating a stronger defensive response known as effector-triggered immunity (ETI) [114]. These immune responses, PTI and ETI, play essential roles in defending against various pathogens and pests. While resistance driven by ETI has been demonstrated to work against nematodes, the effectiveness of PTI in such contexts is not yet established [115 - 117].

The predominant group of resistance genes, commonly known as R-genes, is defined by the presence of a central nucleotide-binding (NB) domain linked to a leucine-rich repeat (LRR) domain on the C-terminal end. Variations do occur in the N-terminal sections across different NB-LRR proteins, but the LRR domain is a frequent element among most R-genes [118]. The LRR domain's extensive variability is believed to enable the recognition of various pathogen-derived molecules, a feature likely facilitated by point mutations under positive selective pressures [119]. Proteins with N-terminal regions resembling the Toll and interleukin-1 receptor (TIR) are classified as TIR-NB-LRRs (TNLs). Conversely, those that do not feature a TIR domain but include a coiled-coil (CC) motif, and sometimes a DNA-binding domain such as the BED Zinc finger or solanaceous domains, are referred to as CC-NB-LRRs (CNLs). This classification has been supported by various studies [119 - 122].

Nematodes significantly impact various crops, making it essential to develop crop varieties that are resistant to these pests. Research has focused on identifying and understanding nematode resistance genes (Nem-R-genes), which include both dominant single genes and key resistance genes similar to those that protect plants from other pathogens [123]. Studies on important quantitative trait loci like Rhg1 and Rhg4, which confer resistance to soybean CNs (SCNs), suggest that protection against PPNs might involve more complex processes than just effector-triggered immunity [124 - 126].

The discovery of the Hs1Pro-1 gene from wild sugar beets marked a significant advancement in the fight against the sugar beet CN (SBCNs), *H. schachtii* [127]. This gene's exact mode of action in providing resistance remains unclear, partly because its protein lacks the leucine-rich repeat (LRR) domain commonly found in other resistance genes. Among these are Mi-1 and Hero A in tomatoes, Gpa2 and Gro1-4 in potatoes, Ma in plums, and the rhg genes in soybean [128 - 132]. Specifically, the tomato gene Mi1.2 triggers a hypersensitive response that kills nearby cells when RKNs attempt to establish an NFS [128, 130]. Meanwhile, Hero A in tomatoes activates a hypersensitive reaction around new NFCS formed by PCNs [128], which differ from typical resistance genes; Rhg1 involves multiple tandem repeats including diverse genes, and Rhg4 codes for a serine hydroxymethyl transferase [133]. However, comprehensive elucidation of the underlying mechanisms governing resistance remains elusive.

The substantial documentation of microRNAs (miRNAs) in modulating gene expression pertaining to plant-nematode interactions is well established. MiRNAs, succinct non-coding RNA molecules typically spanning 20–22 nucleotides, are synthesized *via* dicer-like proteins. The molecules originate from MIR genes and are first transcribed by RNA polymerase II into primary RNA transcripts (pri-miRNA) within the nucleus [134, 135]. These pri-miRNAs form stem-loop hairpin structures, known as pre-miRNAs, which are then cleaved by Dicer-like enzymes in the cytoplasm to produce miRNA duplexes [136, 137]. Within the complex known as the RNA-induced silencing complex (RISC), one strand of the miRNA duplex is incorporated and functions either to degrade specific mRNA molecules or inhibit their translation into proteins, facilitated by the argonaute protein AGO1 [138].

Studies have elucidated the implication of endogenous miRNAs from the host in modulating the expression of genes associated with nematode parasitism. For example, in *Arabidopsis* infected with CN (*Heterodera schachtii*), there is differential regulation of 19 miRNA families at various time points, suggesting their significance in host-nematode interactions [139]. Typically, these miRNAs are suppressed in the initial phase of infection but tend to be overexpressed as the infection progresses. In another study on *Arabidopsis* interacting with *Meloidogyne javanica*, a reduction in miRNA levels in the gall tissues during the early stages of infection has also been observed [140, 141].

Breeding for Nematode Resistance

Breeding for nematode resistance is a crucial aspect of plant genetics aimed at developing crops that can withstand nematode infestations. This process involves traditional breeding techniques combined with modern genomic tools to enhance resistance traits in plants. Breeders meticulously curate progenitor flora exhibiting advantageous phenotypes, notably resilience against nematodes, and intricately interbreed them to yield progeny manifesting heightened resistance. Through successive generations of selection and hybridization, breeders create new varieties that exhibit enhanced resistance to nematodes.

Numerous techniques have been used historically to enhance resistance in plants against nematodes through genetic engineering. Successful examples include the transfer of resistance genes such as the Mi gene from tomatoes against *M. incognita*, the Hs1pro-1 gene from sugar beets for *H. schachtii*, Gpa-2 from potatoes against *G. pallida*, and the Hero A gene from tomatoes for battling *G. rostochiensis* [142, 143]. Additionally, increasing the production of protease inhibitors like the cowpea trypsin inhibitor (CpTI), PIN2, cystatins, and serine proteases has proven effective in developing nematode-resistant plant varieties [144]. A crucial strategy also includes targeting vital nematode effectors through RNA interference (RNAi) technology to curb nematode infestations [144, 145]. Recent studies suggest that altering gene expression within syncytia could further amplify resistance against nematodes in plants.

The modulation of gene expression in plants holds the potential for inducing resistance against CNs. One approach entails genetically modifying plants to either amplify genes linked to resistance or suppress genes vital for syncytium development and nematode viability [145]. Accurate delivery of transgenes into feeding sites is essential, as constant amplification or suppression of specific genes may adversely affect the overall health and growth of the plant [58, 145].

Promoters that are specifically active in areas affected by PPN, such as Pdf2.1 or MIOX5, are ideal for this purpose [146]. Utilizing such promoters enables targeted gene expression at these critical sites, minimizing unwanted effects on the plant's normal physiological functions. Additionally, promoters like Pdf2.1 or MIOX5, which are highly active in syncytia, allow for precise genetic interventions to disrupt nematode development [146, 147]. Other specialized promoters that activate in roots and root tips have also been applied to direct the expression of proteinase inhibitors and peptides that disrupt nematode signaling, across different plant species [148 - 150].

Analyses of transcriptomes across a spectrum of plant species infected with diverse strains of nematodes consistently reveal an elevation in the expression of genes pivotal for the formation of NFS in the roots. Genetic studies employing knockout methodologies aimed at disrupting two specific endo-1,4-β-glucanases, which exhibited significant upregulation within these feeding structures, demonstrated a diminished susceptibility of *Arabidopsis* plants to beet CNs [151]. Likewise, the activation of an ATPase gene (At1g64110) within syncytia induced by *Heterodera schachtii* has been documented [58]. Utilizing promoters tailored for syncytia, like prom.AtPDF2.1 and prom.AtMIOX5, led to a notable reduction in nematode populations when used to significantly lower the expression of this gene.

Nematodes demonstrate a remarkable capacity to attenuate plant defense mechanisms, as evidenced by the suppression of genes associated with defense pathways within feeding sites, as revealed by plant transcriptome analyses following the nematode attack [103, 152 - 154]. Significantly, there was a marked decline observed in the peroxidase gene cluster, where 14 peroxidases ranked among the foremost 100 genes showing the most notable decrease in expression levels [154]. The transcription factor AtRAP2.6, which is responsive to ethylene and found in *Arabidopsis*, was observed to be markedly reduced in expression within syncytia. Interestingly, this gene was previously overexpressed using a ubiquitous promoter [58]. These findings further the debate about the suitability of using the CaMV35S promoter for gene expression in *Arabidopsis* root syncytia caused by beet CNs.

Research has shown that targeting genetic pathways responsible for producing protective compounds such as camalexin and callose can boost plant defenses [155 - 158]. Specifically, the use of the FMV-Sgt promoter to drive the expression of AtPAD4 in soybeans significantly increases resistance to SCNs and RKNs by reducing mature *H. glycines* females and *M. incognita* galls by as much as 68% and 77%, respectively [159]. Additional studies have explored the camalexin synthesis pathway in the context of plant-nematode interactions, highlighting how

AtWRKY33 and AtPAD3 mutations or over-expressions affect plant responses [147]. In particular, the heightened expression of WRKY33 in *Arabidopsis* syncytia has been linked to reduced nematode susceptibility. Furthermore, enhancing the expression of the GmSAMT1 gene, which is involved in salicylic acid methylation in soybeans, has proven effective in conferring SCN resistance [160]. Developing nematode-resistant plant varieties is essential for promoting sustainable and productive agriculture, reducing reliance on chemical nematicides, minimizing crop losses due to nematodes, and supporting eco-friendly agricultural practices.

MANAGEMENT STRATEGIES FOR NEMATODE CONTROL

Plant parasitic nematodes present considerable obstacles to global agriculture, resulting in substantial decreases in crop yields and economic harm each year. These tiny roundworms compromise plant vitality and output by feeding on plant roots, disrupting both nutrient absorption and water intake. It is crucial to employ effective tactics to handle nematode-related losses and ensure the sustainable production of crops. Below, we outline the strategies for managing plant parasitic nematodes and stress the significance of adopting comprehensive management methodologies to minimize nematode-related harm and maximize crop yields. This endeavor is critical for safeguarding both food security and economic stability within agricultural systems.

Cultural Practices

Sanitation

The spread of PPNs to agricultural lands can occur through multiple routes, including infected farming tools, plant materials, water sources used for irrigation, and human activities [161, 162]. To curb the proliferation of PPNs, several preventative tactics can be implemented. These include the utilization of sterilized agricultural instruments during and after each crop cycle, ensuring footwear does not carry soil harboring nematodes, and adhering to strict sanitation protocols. Such protocols may involve the extraction and destruction of affected plants and the incineration of plant debris and weeds. It is also critical to prevent the transfer of plants and soil from areas known to have PPN occurrences [162 - 164]. Incorporating strategies like pre-season tillage, meticulous weed management, and plant desiccation prior to sowing have shown success in reducing PPN densities in crops [164, 165]. Additionally, it is advisable for growers to plant only nematode-free seeds or seedlings and to use uncontaminated water for both irrigation and fertilization to further diminish the threat of nematode infection [166].

Crop Rotation

Rotating crops is a crucial strategy for managing plant-parasitic nematodes (PPNs), which works by alternating between crops that are vulnerable to nematodes and those that resist them, thus interrupting the nematodes' reproductive cycle. This tactic not only starves the nematode population by withholding their preferred hosts but also diminishes their numbers in the soil, enhancing overall plant health and agricultural output [167 - 169]. The advantages extend beyond nematode control, contributing to enhanced soil fertility and structure, thus promoting better growth and increased yields of subsequent crops [170 - 175]. However, when choosing crops for rotation, it is important to balance nematode resistance with economic factors to ensure that the crop grown after the rotation period can compensate for the use of the land during the rotation phase [176 - 180]. Marigold (*Tagetes* spp.) is particularly noted for its effectiveness against a range of PPN species and its economic value as a floral product, making it a prime candidate for rotation [181, 182]. It is also vital to perform frequent soil assessments to monitor the types of nematodes present, as this information is crucial for selecting the most appropriate rotation crops tailored to the specific nematode threats in a field [183].

Cover Crops for PPN Management

Incorporating cover crops into agricultural practices not only promotes soil health but also effectively diminishes populations of PPNs in fields planted with susceptible crops [170, 181, 184, 185]. Certain cover crops exert toxic effects on PPNs and foster beneficial soil microbes that combat these nematodes and enhance soil structure, thereby reducing the impact of nematodes on crops [170, 186 - 189]. Biofumigant plants such as Brassicas, Crotalaria, sorghum, and Sudan grass release compounds that mirror the nematocidal properties of methyl bromide, offering effective control of various PPN species when grown as cover crops or used as green manure [182, 190]. Research also shows that cultivating soybean varieties resistant to *Rotylenchulus reniformis* as a cover crop in cotton fields can suppress this nematode to levels that are not economically damaging to subsequent cotton crops susceptible to the same parasite [177, 191].

Organic Amendments

Incorporating organic substances such as compost and manure into the soil not only enhances its fertility but also boosts the diversity and activity of beneficial microbes. These microbes play a crucial role in reducing populations of harmful nematodes. Research indicates that the use of materials like poultry and cow dung, along with household waste and synthetic fertilizers, has been effective in curtailing nematode levels in crops like Ethiopian eggplants [192]. Similarly, studies have shown that the application of poultry litter and rapeseed can decrease the number of nematode eggs, root galling, and overall nematode numbers in tomatoes [193]. Introducing 20 tons per hectare of compost and farmyard manure has also been proven to significantly diminish plant-parasitic nematode (PPN) populations [194]. Moreover, Rosskopf *et al.* (2020) demonstrated that compost enhances the proliferation of predatory mites that prey on nematodes, while manure applications boost microbial activities that are harmful to PPNs due to increased nutrient availability [195]. Additionally, the use of green manure, as well as poultry and cattle manure mixed with oilseed cake, has been shown to suppress nematode populations by encouraging the growth of microbes that antagonize them and by emitting toxic substances [196]. Implementing strategies that improve soil health through organic amendments is a viable approach to managing PPNs, especially in settings where commercial biological agents are either costly or not readily available. This method is particularly feasible for small-scale farmers and in controlled agricultural environments [197].

Chemical Measures

Chemical pesticides have historically played a crucial role in agricultural pest management, significantly enhancing crop yields. This includes both fumigant and non-fumigant pesticides utilized for controlling PPNs and other soil pathogens. Methyl bromide, a widely utilized all-purpose fumigant pesticide for combating PPNs, was discontinued in 2005 [168]. Most nematicidal fumigants have been phased out in agricultural practices due to their harmful environmental impact, leaving only a few like chloropicrin, 1,3-dichloropropene, and dazomet in use [198, 199]. As an alternative to the banned methyl bromide, metham sodium has been applied *via* drip irrigation, demonstrating its effectiveness [200, 201]. In polyhouses, using a mix of Oxamyl, metham sodium, and 1,3-dichloropropene has proven successful in managing PPNs and boosting crop yields [202].

To date, novel nematicides like cyclobutrifluram, fluensulfone, fluopyram, and fluazaindolizine are garnering interest among growers and have had favor because of their low toxicity and high level of efficacy against PPNs [199, 203 - 209]. Fluensulfone is available in various formulations and is registered for multiple crops to offer protection against various PPN species [199]. Another compound, fluopyram, targets nematode mitochondrial respiration and is utilized in diverse agricultural settings, including greenhouses, to manage various nematode species like *Pratylenchus* spp., *Longidorus* spp., and *Meloidogyne* spp [199, 210]. Similarly, fluazaindolizine exhibits efficacy against *Meloidogyne* spp., though no commercial product is available [204]. Although considered a better fungicide than a nematicide, Chloropicrin as a fumigant has a considerable level of control efficacy control against RKN species [211, 212]. Despite their potential, non-fumigant nematicides like oxamyl, chloropicrin, fenamiphos, and aldicarb sometimes show variable results against nematodes, influenced by the specific PPN species, the crop, and environmental conditions [211 - 214]. In light of the global agricultural shift towards sustainability, chemical applications should be minimized or integrated into holistic approaches.

Biological Control

Biological methods of controlling nematodes provide a sustainable option for reducing dependence on chemical treatments in agriculture. Over recent years, significant progress has been made in the use of natural predators such as fungi, bacteria, and other microbes to decrease the damage caused by nematodes to crops, thus supporting more environmentally friendly agricultural practices and enhancing food security [215 - 220]. Among these, nematophagous fungi have been particularly effective in combating plant-parasitic nematodes (PPNs). Effective species include *Duddingtonia flagrans*, *Dactylaria* spp., *Arthrobotrys* spp., and *Monacrosporium* spp [221 - 225]. These fungi naturally occur in the soil around plant roots, where they provide a defensive barrier against PPNs [226, 227]. Their unique methods of attacking nematodes differentiate them from other microbial agents such as bacteria and viruses [223 - 225]. Fungi that trap nematodes through adhesive and mechanical means have also been shown to be effective, with notable successes in species like *Arythrobotrys oligospora*, *Candelabrella musiformis*, and *Dactylaria eudermata* used against *M. incognita* in tobacco [228, 229], and *Arthrobotrys oligospora* and *Meria coniospora* used against the same nematode and other RKNs in tomatoes [230, 231]. Some nematophagous fungi possess the ability to penetrate the eggs of phytoparasitic nematodes, destroying them through enzymatic actions on the eggshells [232, 233]. Notable among these are *Pochonia chlamydosporia* and *Purpureocillium lilacinum*, which target *M. incognita*, and *Paecilomyces lilacinus*, used against *Meloidogyne* spp. in tomatoes [233, 234]. Additionally, fungi like *Trichoderma* spp. have a long history of being utilized in commercial products worldwide for controlling PPNs and other plant pathogens, demonstrating significant global efficacy [235 - 239].

Several bacterial species have demonstrated effectiveness in managing PPNs, making them viable biocontrol agents in agricultural settings. Research shows that *B. subtilis* can combat root lesion nematodes and RKNs in sugarcane and common beans [240, 241]. Similarly, *B. aryabhattai* has shown promising results in decreasing CN populations in soybean crops [242]. Other studies reveal that *B. cereus* and *B. megaterium* can control RKN *M. javanica* in eggplant and cucumber fields, respectively [243]. Laboratory and controlled environment studies further underscore the potential of bacterial agents against PPNs [244].

observed that *B. pumilus* significantly hindered the survival and egg hatching of *M. arenaria*, noting up to 90% mortality and 88% inhibition of egg hatch. Additionally, research by [245] indicated that strains of *B. subtilis, B. pumilus, B. megaterium,* and *P. fluorescens* significantly reduced egg masses and galls of *M. incognita* in sugar beet roots, achieving control efficacy exceeding 60% [246]. reported a 67.1% reduction in *M. incognita* attacks on tomato roots by *B. cereus*, whose secretions also deterred nematodes in the rhizosphere. In greenhouse settings, P. *fluorescens* and *S. marcescens* markedly diminished nematode galls and egg masses of *M. incognita* on cucumbers [247]. Furthermore, *Xenorhabdus* and *Photorhabdus* bacteria suspensions effectively reduced galls, egg masses, and reproduction rates of Northern RKN, *Meloidogyne hapla* [248], and *Xenorhabdus bovienii* was noted for its efficacy against RKNs [249]. These collective findings highlight the strong potential of bacterial biocontrol agents against PPNs, suggesting a need for more extensive field trials to validate their effectiveness in real-world agricultural conditions.

CONCLUSION AND FUTURE PERSPECTIVES IN PLANT PARASITIC NEMATODE MANAGEMENT

PPNs are a major agricultural challenge, leading to significant crop damage and loss of yield on par with other pests and diseases. Often, the presence and extent of damage caused by these nematodes go unrecognized due to identification challenges. Multiple management approaches are adopted to mitigate the economic impacts of PPNs on crops. These methods include using nematicides, practicing specific agricultural techniques, and breeding crops that are resistant to nematodes.

Environmental concerns limit the application of nematicides. Cultural management strategies, such as crop rotation, face obstacles because CNs can lie dormant for long periods, and RKNs can infest a wide range of plants. Therefore, developing resistant crop varieties is considered the most effective defense, achieved either through natural adaptation or by introducing resistant genes from wild relatives through traditional breeding methods. Traditional breeding, which involves crossbreeding and selection, is a slow process that also narrows the genetic diversity. While gene transfer techniques have successfully introduced resistance traits among closely related species, they often fall short when applied to genetically distant species.

The exploration of molecular interactions between plants and nematodes is increasingly recognized as crucial. A deeper understanding of these interactions could lead to the development of targeted, environmentally benign measures to control these pests. Such innovations are key to decreasing the substantial economic losses that nematodes inflict on the agricultural sector.

REFERENCES

[1]   Bharti L, Bhat AH, Chaubey AK, Abolafia J. Morphological and molecular characterisation of *Merlinius brevidens* (Allen, 1955) Siddiqi, 1970 (Nematoda: Rhabditida: Merlinidae) from India. J Nat Hist 2020; 54(23-24): 1477-98.
[http://dx.doi.org/10.1080/00222933.2020.1810352]

[2]   D Howland A, Quintanilla M. Plant-Parasitic nematodes and their effects on ornamental plants: A review. J Nematol 2023; 55(1): 20230007.
[http://dx.doi.org/10.2478/jofnem-2023-0007] [PMID: 37082221]

[3]   Elling AA. Major emerging problems with minor *Meloidogyne* species. Phytopathology 2013; 103(11): 1092-102.
[http://dx.doi.org/10.1094/PHYTO-01-13-0019-RVW] [PMID: 23777404]

[4]   Phani V, Khan MR, Dutta TK. Plant-parasitic nematodes as a potential threat to protected agriculture: Current status and management options. Crop Prot 2021; 144: 105573.
[http://dx.doi.org/10.1016/j.cropro.2021.105573]

[5]   Bernard GC, Egnin M, Mortley D, Bonsi C. Nematode problems in bulb crops and sustainable management. In: Khan MR, Quintanill M, Eds. Nematode diseases of crops and their sustainable management. Elsevier 2023; pp. 297-309.
[http://dx.doi.org/10.1016/B978-0-323-91226-6.00006-7]

[6]   Blok VC, Jones JT, Phillips MS, Trudgill DL. Parasitism genes and host range disparities in biotrophic nematodes: The conundrum of polyphagy *versus* specialisation. BioEssays 2008; 30(3): 249-59.
[http://dx.doi.org/10.1002/bies.20717] [PMID: 18293363]

[7]   Levin KA, Tucker MR, Strock CF, Lynch JP, Mather DE. Three-dimensional imaging reveals that positions of cyst nematode feeding sites relative to xylem vessels differ between susceptible and resistant wheat. Plant Cell Rep 2021; 40(2): 393-403.
[http://dx.doi.org/10.1007/s00299-020-02641-w] [PMID: 33388893]

[8]   Vanholme B, De Meutter J, Tytgat T, Van Montagu M, Coomans A, Gheysen G. Secretions of plant-parasitic nematodes: A molecular update. Gene 2004; 332: 13-27.
[http://dx.doi.org/10.1016/j.gene.2004.02.024] [PMID: 15145050]

[9]   Favery B, Quentin M, Jaubert-Possamai S, Abad P. Gall-forming root-knot nematodes hijack key plant cellular functions to induce multinucleate and hypertrophied feeding cells. J Insect Physiol 2016; 84: 60-9.
[http://dx.doi.org/10.1016/j.jinsphys.2015.07.013] [PMID: 26211599]

[10]   Kyndt T, Vieira P, Gheysen G, de Almeida-Engler J. Nematode feeding sites: Unique organs in plant roots. Planta 2013; 238(5): 807-18.
[http://dx.doi.org/10.1007/s00425-013-1923-z] [PMID: 23824525]

[11]   Palomares-Rius JE, Escobar C, Cabrera J, Vovlas A, Castillo P. Anatomical alterations in plant tissues induced by plant-parasitic nematodes. Front Plant Sci 2017; 8: 1987.
[http://dx.doi.org/10.3389/fpls.2017.01987] [PMID: 29201038]

[12]   Rodiuc N, Vieira P, Banora MY, de Almeida Engler J. On the track of transfer cell formation by specialized plant-parasitic nematodes. Front Plant Sci 2014; 5: 160.
[http://dx.doi.org/10.3389/fpls.2014.00160] [PMID: 24847336]

[13]   Nguyen CN, Perfus-Barbeoch L, Quentin M, *et al.* A root-knot nematode small glycine and cysteine-rich secreted effector, MiSGCR1, is involved in plant parasitism. New Phytol 2018; 217(2): 687-99.
[http://dx.doi.org/10.1111/nph.14837] [PMID: 29034957]

[14]   Ali MA, Anjam MS, Nawaz MA, Lam HM, Chung G. Signal transduction in plant–nematode interactions. Int J Mol Sci 2018; 19(6): 1648.
[http://dx.doi.org/10.3390/ijms19061648] [PMID: 29865232]

[15]   Hewezi T. Cellular signaling pathways and posttranslational modifications mediated by nematode effector proteins. Plant Physiol 2015; 169(2): 1018-26.
[http://dx.doi.org/10.1104/pp.15.00923] [PMID: 26315856]

[16]   Hewezi T, Baum TJ. Manipulation of plant cells by cyst and root-knot nematode effectors. Mol Plant Microbe Interact 2013; 26(1): 9-16.
[http://dx.doi.org/10.1094/MPMI-05-12-0106-FI] [PMID: 22809272]

[17]   Mitchum MG, Hussey RS, Baum TJ, *et al.* Nematode effector proteins: An emerging paradigm of parasitism. New Phytol 2013; 199(4): 879-94.
[http://dx.doi.org/10.1111/nph.12323] [PMID: 23691972]

[18]   Quentin M, Abad P, Favery B. Plant parasitic nematode effectors target host defense and nuclear functions to establish feeding cells. Front Plant Sci 2013; 4: 53.
[http://dx.doi.org/10.3389/fpls.2013.00053] [PMID: 23493679]

[19]   Vieira P, Gleason C. Plant-parasitic nematode effectors — insights into their diversity and new tools for their identification. Curr Opin Plant Biol 2019; 50: 37-43.
[http://dx.doi.org/10.1016/j.pbi.2019.02.007] [PMID: 30921686]

[20]   Siddique S, Grundler FMW. Parasitic nematodes manipulate plant development to establish feeding sites. Curr Opin Microbiol 2018; 46: 102-8.
[http://dx.doi.org/10.1016/j.mib.2018.09.004] [PMID: 30326406]

[21]   Ali MA, Azeem F, Li H, Bohlmann H. Smart parasitic nematodes use multifaceted strategies to parasitize plants. Front Plant Sci 2017; 8: 1699.
[http://dx.doi.org/10.3389/fpls.2017.01699] [PMID: 29046680]

[22]   Bellafiore S, Briggs SP. Nematode effectors and plant responses to infection. Curr Opin Plant Biol 2010; 13(4): 442-8.
[http://dx.doi.org/10.1016/j.pbi.2010.05.006] [PMID: 20542724]

[23]   Gheysen G, Mitchum MG. How nematodes manipulate plant development pathways for infection. Curr Opin Plant Biol 2011; 14(4): 415-21.
[http://dx.doi.org/10.1016/j.pbi.2011.03.012] [PMID: 21458361]

[24]   Gheysen G, Fenoll C. Gene expression in nematode feeding sites. Annu Rev Phytopathol 2002; 40(1): 191-219.
[http://dx.doi.org/10.1146/annurev.phyto.40.121201.093719] [PMID: 12147759]

[25]   Lozano-Torres JL, Wilbers RHP, Gawronski P, *et al.* Dual disease resistance mediated by the immune receptor Cf-2 in tomato requires a common virulence target of a fungus and a nematode. Proc Natl Acad Sci USA 2012; 109(25): 10119-24.
[http://dx.doi.org/10.1073/pnas.1202867109] [PMID: 22675118]

[26]   Sobczak M, Avrova A, Jupowicz J, Phillips MS, Ernst K, Kumar A. Characterization of susceptibility and resistance responses to potato cyst nematode (*Globodera* spp.) infection of tomato lines in the absence and presence of the broad-spectrum nematode resistance Hero gene. Mol Plant Microbe Interact 2005; 18(2): 158-68.
[http://dx.doi.org/10.1094/MPMI-18-0158] [PMID: 15720085]

[27]   Sobczak M, Golinowski W, Grundler FMW. Ultrastructure of feeding plugs and feeding tubes formed by *Heterodera schachtii*. Nematology 1999; 1(4): 363-74.
[http://dx.doi.org/10.1163/156854199508351]

[28]   Grundler FMW, Sobczak M, Golinowski W. Formation of wall openings in root cells of *Arabidopsis thaliana* following infection by the plant-parasitic nematode *Heterodera schachtii*. Eur J Plant Pathol 1998; 104(6): 545-51.

[http://dx.doi.org/10.1023/A:1008692022279]

[29]   Grundler FMW, Wyss U. Seminar: *Heterodera schachtii* and *Arabidopsis thaliana*, a model host-parasite interaction. Nematologica 1992; 38(1-4): 488-93.
[http://dx.doi.org/10.1163/187529292X00450]

[30]   Sato K, Kadota Y, Shirasu K. Plant immune responses to parasitic nematodes. Front Plant Sci 2019; 10: 1165.
[http://dx.doi.org/10.3389/fpls.2019.01165] [PMID: 31616453]

[31]   Teillet A, Dybal K, Kerry BR, Miller AJ, Curtis RHC, Hedden P. Transcriptional changes of the root-knot nematode *Meloidogyne incognita* in response to *Arabidopsis thaliana* root signals. Ghanim M, Ed. PLoS ONE. 2013;8:e61259.
[http://dx.doi.org/10.1371/journal.pone.0061259]

[32]   Sobczak M, Golinowski W. Cyst nematodes and syncytia. In: Jones J, Gheysen G, Fenoll C, Eds. Genomics and molecular genetics of plant-nematode interactions. Dordrecht: Springer Science+Business Media BV; 2011. p. 61–82.
[http://dx.doi.org/10.1007/978-94-007-0434-3]

[33]   Holbein J, Franke RB, Marhavý P, *et al.* Root endodermal barrier system contributes to defence against plant-parasitic cyst and root-knot nematodes. Plant J 2019; 100(2): 221-36.
[http://dx.doi.org/10.1111/tpj.14459] [PMID: 31322300]

[34]   Jones MGK, Goto DB. Root-knot nematodes and giant cells. In: Jones J, Gheysen G, Fenoll C, Eds. Genomics and molecular genetics of plant-nematode interactions. Dordrecht: Springer Netherlands 2011; pp. 83-100.
[http://dx.doi.org/10.1007/978-94-007-0434-3_5]

[35]   Price JA, Coyne D, Blok VC, Jones JT. Potato cyst nematodes *Globodera rostochiensis* and *G. pallida*. Mol Plant Pathol 2021; 22(5): 495-507.
[http://dx.doi.org/10.1111/mpp.13047] [PMID: 33709540]

[36]   Castagnone-Sereno P. Genetic variability and adaptive evolution in parthenogenetic root-knot nematodes. Heredity 2006; 96(4): 282-9.
[http://dx.doi.org/10.1038/sj.hdy.6800794] [PMID: 16404412]

[37]   Abad P, Williamson VM. Plant nematode interaction: A sophisticated dialogue. Adv Bot Res. Elsevier; 2010 . p. 147–192. Available from: https://linkinghub.elsevier.com/retrieve/pii/S0065229610530052

[38]   Behera SK, Sahu A. Disease complex: Nematode and plant pathogen interaction. Singh HK, Ed, Current research and innovations in plant pathology 2022; 109-44.
[http://dx.doi.org/10.22271/ed.book.1874]

[39]   Jones MGK, Payne HL. The structure of syncytia induced by the phytoparasitic nematode *Nacobbus aberrans* in tomato roots, and the possible role of plasmodesmata in their nutrition. J Cell Sci 1977; 23(1): 299-313.
[http://dx.doi.org/10.1242/jcs.23.1.299] [PMID: 197113]

[40]   Chen Q, Rehman S, Smant G, Jones JT. Functional analysis of pathogenicity proteins of the potato cyst nematode *Globodera rostochiensis* using RNAi. Mol Plant Microbe Interact 2005; 18(7): 621-5.
[http://dx.doi.org/10.1094/MPMI-18-0621] [PMID: 16042007]

[41]   Shibuya H, Kikuchi T. Purification and characterization of recombinant endoglucanases from the pine wood nematode *Bursaphelenchus xylophilus*. Biosci Biotechnol Biochem 2008; 72(5): 1325-32.
[http://dx.doi.org/10.1271/bbb.70819] [PMID: 18460801]

[42]   Smant G, Stokkermans JPWG, Yan Y, *et al.* Endogenous cellulases in animals: Isolation of β-1, 4-endoglucanase genes from two species of plant-parasitic cyst nematodes. Proc Natl Acad Sci USA 1998; 95(9): 4906-11.
[http://dx.doi.org/10.1073/pnas.95.9.4906] [PMID: 9560201]

[43]   Kikuchi T, Furlanetto C, Jones J. Horizontal gene transfer from bacteria and fungi as a driving force in the evolution of plant parasitism in nematodes. Nematology 2005; 7(5): 641-6.
[http://dx.doi.org/10.1163/156854105775142919]

[44]   Rybarczyk-Mydłowska K, Maboreke HR, van Megen H, *et al.* Rather than by direct acquisition *via* lateral gene transfer, GHF5 cellulases were passed on from early Pratylenchidae to root-knot and cyst nematodes. BMC Evol Biol 2012; 12(1): 221.
[http://dx.doi.org/10.1186/1471-2148-12-221] [PMID: 23171084]

[45]   Hewezi T, Howe P, Maier TR, *et al.* Cellulose binding protein from the parasitic nematode *Heterodera schachtii* interacts with *Arabidopsis* pectin methylesterase: Cooperative cell wall modification during parasitism. Plant Cell 2008; 20(11): 3080-93.
[http://dx.doi.org/10.1105/tpc.108.063065] [PMID: 19001564]

[46]   Haegeman A, Mantelin S, Jones JT, Gheysen G. Functional roles of effectors of plant-parasitic nematodes. Gene 2012; 492(1): 19-31.
[http://dx.doi.org/10.1016/j.gene.2011.10.040] [PMID: 22062000]

[47]   Durner J, Shah J, Klessig DF. Salicylic acid and disease resistance in plants. Trends Plant Sci 1997; 2(7): 266-74.
[http://dx.doi.org/10.1016/S1360-1385(97)86349-2]

[48]   Bauters L, Kyndt T, De Meyer T, *et al.* Chorismate mutase and isochorismatase, two potential effectors of the migratory nematode *Hirschmanniella oryzae*, increase host susceptibility by manipulating secondary metabolite content of rice. Mol Plant Pathol 2020; 21(12): 1634-46.
[http://dx.doi.org/10.1111/mpp.13003] [PMID: 33084136]

[49]   Iberkleid I, Sela N, Brown Miyara S. *Meloidogyne javanica* fatty acid- and retinol-binding protein (Mj-FAR-1) regulates expression of lipid-, cell wall-, stress- and phenylpropanoid-related genes during nematode infection of tomato. BMC Genomics 2015; 16(1): 272.
[http://dx.doi.org/10.1186/s12864-015-1426-3] [PMID: 25886179]

[50]   Baum TJ, Hussey RS, Davis EL. Root-knot and cyst nematode parasitism genes: The molecular basis of plant parasitism. In: Setlow JK, Ed. Genetic engineering. Springer US 2007; pp. 17-43.
[http://dx.doi.org/10.1007/978-0-387-34504-8_2]

[51]   Andres MF, Melillo MT, Delibes A, Romero MD, Bleve-Zacheo T. Changes in wheat root enzymes correlated with resistance to cereal cyst nematodes. New Phytol 2001; 152(2): 343-54.
[http://dx.doi.org/10.1111/j.0028-646X.2001.00258.x]

[52]   Balhadère P, Evans AAF. Cytochemical investigation of resistance to root-knot nematode *Meloidogyne naasi* in cereals and grasses using cryosections of roots. Fundam Appl Nematol 1995; 18: 539-47.

[53]   Dhakshinamoorthy S, Mariama K, Elsen A, De Waele D. Phenols and lignin are involved in the defence response of banana (*Musa*) plants to *Radopholus similis* infection. Nematology 2014; 16(5): 565-76.
[http://dx.doi.org/10.1163/15685411-00002788]

[54]   Wuyts N, Lognay G, Swennen R, De Waele D. Nematode infection and reproduction in transgenic and mutant *Arabidopsis* and tobacco with an altered phenylpropanoid metabolism. J Exp Bot 2006; 57(11): 2825-35.
[http://dx.doi.org/10.1093/jxb/erl044] [PMID: 16831845]

[55]   Ji H, Kyndt T, He W, Vanholme B, Gheysen G. β-Aminobutyric Acid–Induced Resistance against root-knot nematodes in rice is based on increased basal defense. Mol Plant Microbe Interact 2015; 28(5): 519-33.
[http://dx.doi.org/10.1094/MPMI-09-14-0260-R] [PMID: 25608179]

[56]   Huang WK, Ji HL, Gheysen G, Kyndt T. Thiamine-induced priming against root-knot nematode infection in rice involves lignification and hydrogen peroxide generation. Mol Plant Pathol 2016;

17(4): 614-24.
[http://dx.doi.org/10.1111/mpp.12316] [PMID: 27103216]

[57]   Fujimoto T, Mizukubo T, Abe H, Seo S. Sclareol induces plant resistance to root-knot nematode partially through ethylene-dependent enhancement of lignin accumulation. Mol Plant Microbe Interact 2015; 28(4): 398-407.
[http://dx.doi.org/10.1094/MPMI-10-14-0320-R] [PMID: 25423264]

[58]   Ali MA, Abbas A, Kreil DP, Bohlmann H. Overexpression of the transcription factor RAP2.6 leads to enhanced callose deposition in syncytia and enhanced resistance against the beet cyst nematode *Heterodera schachtii* in *Arabidopsis* roots. BMC Plant Biol 2013; 13(1): 47.
[http://dx.doi.org/10.1186/1471-2229-13-47] [PMID: 23510309]

[59]   Abad P, Castagnone-Sereno P, Rosso MN, Engler JDA, Favery B. Invasion, feeding and development. In: Perry RN, Moens M, Starr JL, Eds. Root-Knot Nematodes. 1ˢᵗ ed. UK: CABI 2009; pp. 163-81.
[http://dx.doi.org/10.1079/9781845934927.0163]

[60]   Grundler FMW, Munch A, Wyss U. The Parasitic behaviour of second-stage juveniles of *Meloidogyne incognita* in roots of *Arabidopsis thaliana.* Nematologica 1992; 38(1-4): 98-111.
[http://dx.doi.org/10.1163/187529292X00081]

[61]   Hung C, Rohde RA. Phenol accumulation related to resistance in tomato to infection by root-knot and lesion nematodes. J Nematol 1973; 5(4): 253-8.
[PMID: 19319346]

[62]   Pegard A, Brizzard G, Fazari A, Soucaze O, Abad P, Djian-Caporalino C. Histological characterization of resistance to different root-knot nematode species related to phenolics accumulation in *Capsicum annuum*. Phytopathology 2005; 95(2): 158-65.
[http://dx.doi.org/10.1094/PHYTO-95-0158] [PMID: 18943985]

[63]   Knypl JS, Chylinska KM, Brzeski MW. Increased level of chlorogenic acid and inhibitors of indolyl-3-acetic acid oxidase in roots of carrot infested with the northern root-knot nematode. Physiol Plant Pathol 1975; 6(1): 51-64.
[http://dx.doi.org/10.1016/0048-4059(75)90104-6]

[64]   Gill JR, Harbornez JB, Plowright RA, Grayer RJ, Rahman ML. The Induction of phenolic compounds in rice after infection by the stem nematode *Ditylenchus angustus.* Nematologica 1996; 42(5): 564-78.
[http://dx.doi.org/10.1163/004625996X00063]

[65]   D'Addabbo T, Carbonara T, Argentieri MP, *et al.* Nematicidal potential of *Artemisia annua* and its main metabolites. Eur J Plant Pathol 2013; 137(2): 295-304.
[http://dx.doi.org/10.1007/s10658-013-0240-5]

[66]   López-Martínez N, Colinas-León MT, Peña-Valdivia CB, *et al.* Alterations in peroxidase activity and phenylpropanoid metabolism induced by *Nacobbus aberrans* Thorne and Allen, 1944 in chilli (*Capsicum annuum* L.) CM334 resistant to *Phytophthora capsici* Leo. Plant Soil 2011; 338(1-2): 399-409.
[http://dx.doi.org/10.1007/s11104-010-0553-5]

[67]   Hölscher D, Dhakshinamoorthy S, Alexandrov T, *et al.* Phenalenone-type phytoalexins mediate resistance of banana plants (*Musa* spp.) to the burrowing nematode *Radopholus similis.* Proc Natl Acad Sci USA 2014; 111(1): 105-10.
[http://dx.doi.org/10.1073/pnas.1314168110] [PMID: 24324151]

[68]   Luis JG, Fletcher WQ, Echeverri F, Abad T, Kishi MP, Perales A. New phenalenone-type phytoalexins from *Musa acuminata* (Colla AAA) Grand Nain. Nat Prod Lett 1995; 6(1): 23-30.
[http://dx.doi.org/10.1080/10575639508044083]

[69]   Luque-Ortega JR, Martínez S, Saugar JM, *et al.* Fungus-elicited metabolites from plants as an enriched source for new leishmanicidal agents: Antifungal phenyl-phenalenone phytoalexins from the banana plant (*Musa acuminata*) target mitochondria of *Leishmania donovani* promastigotes. Antimicrob Agents Chemother 2004; 48(5): 1534-40.

[http://dx.doi.org/10.1128/AAC.48.5.1534-1540.2004] [PMID: 15105102]

[70]   Chin S, Behm CA, Mathesius U. Functions of flavonoids in plant-nematode interactions. Plants 2018; 7(4): 85.
[http://dx.doi.org/10.3390/plants7040085] [PMID: 30326617]

[71]   Wuyts N, Swennen R, De Waele D. Effects of plant phenylpropanoid pathway products and selected terpenoids and alkaloids on the behaviour of the plant-parasitic nematodes *Radopholus similis, Pratylenchus penetrans* and *Meloidogyne incognita.* Nematology 2006; 8(1): 89-101.
[http://dx.doi.org/10.1163/156854106776179953]

[72]   Baldridge GD, O'Neill NR, Samac DA. Alfalfa (*Medicago sativa* L.) resistance to the root-lesion nematode, *Pratylenchus penetrans*: Defense-response gene mRNA and isoflavonoid phytoalexin levels in roots. Plant Mol Biol 1998; 38(6): 999-1010.
[http://dx.doi.org/10.1023/A:1006182908528] [PMID: 9869406]

[73]   Faizi S, Fayyaz S, Bano S, *et al.* Isolation of nematicidal compounds from *Tagetes patula* L. yellow flowers: Structure-activity relationship studies against cyst nematode *Heterodera zeae* infective stage larvae. J Agric Food Chem 2011; 59(17): 9080-93.
[http://dx.doi.org/10.1021/jf201611b] [PMID: 21780738]

[74]   Kaplan DT, Keen NT, Thomason IJ. Association of glyceollin with the incompatible response of soybean roots to *Meloidogyne incognita.* Physiol Plant Pathol 1980; 16(3): 309-18.
[http://dx.doi.org/10.1016/S0048-4059(80)80002-6]

[75]   Kaplan DT, Thomason IJ, Van Gundy SD. Histological study of the compatible and incompatible interaction of soybeans and *Meloidogyne incognita.* J Nematol 1979; 11(4): 338-43.
[PMID: 19300654]

[76]   Sánchez-Sánchez H, Morquecho-Contreras A. Chemical plant defense against herbivores. In: Shields VDC, Ed. Herbivores. InTech; 2017. Available from: http://www.intechopen.com/books/herbivores/chemical-plant-defense-against
[http://dx.doi.org/10.5772/67346]

[77]   Hamaguchi T, Sato K, Vicente CSL, Hasegawa K. Nematicidal actions of the marigold exudate α-terthienyl: Oxidative stress-inducing compound penetrates nematode hypodermis. Biol Open 2019; 8(4): bio.038646.
[http://dx.doi.org/10.1242/bio.038646] [PMID: 30926596]

[78]   Takasugi M, Yachida Y, Anetai M, Masamune T, Kegasawa K. Identification of asparagusic acid as a nematicide occurring naturally in the roots of *Asparagus.* Chem Lett 1975; 4(1): 43-4.
[http://dx.doi.org/10.1246/cl.1975.43]

[79]   Brown PD, Morra MJ. Control of soil-borne plant pests using glucosinolate-containing plants. Adv Agron. Elsevier 1997; 167-231. Available from: https://linkinghub.elsevier.com/retrieve/pii/S0065211308606641

[80]   Zasada IA, Ferris H. Sensitivity of *Meloidogyne javanica* and *Tylenchulus semipenetrans* to isothiocyanates in laboratory assays. Phytopathology 2003; 93(6): 747-50.
[http://dx.doi.org/10.1094/PHYTO.2003.93.6.747] [PMID: 18943062]

[81]   Bak S, Tax FE, Feldmann KA, Galbraith DW, Feyereisen R. CYP83B1, a cytochrome P450 at the metabolic branch point in auxin and indole glucosinolate biosynthesis in *Arabidopsis.* Plant Cell 2001; 13(1): 101-11.
[http://dx.doi.org/10.1105/tpc.13.1.101] [PMID: 11158532]

[82]   Hull AK, Vij R, Celenza JL. *Arabidopsis* cytochrome P450s that catalyze the first step of tryptophan-dependent indole-3-acetic acid biosynthesis. Proc Natl Acad Sci USA 2000; 97(5): 2379-84.
[http://dx.doi.org/10.1073/pnas.040569997] [PMID: 10681464]

[83]   Mikkelsen MD, Hansen CH, Wittstock U, Halkier BA. Cytochrome P450 CYP79B2 from *Arabidopsis* catalyzes the conversion of tryptophan to indole-3-acetaldoxime, a precursor of indole glucosinolates

and indole-3-acetic acid. J Biol Chem 2000; 275(43): 33712-7.
[http://dx.doi.org/10.1074/jbc.M001667200] [PMID: 10922360]

[84]   Mikkelsen MD, Naur P, Halkier BA. *Arabidopsis* mutants in the *C–S* lyase of glucosinolate biosynthesis establish a critical role for indole-3-acetaldoxime in auxin homeostasis. Plant J 2004; 37(5): 770-7.
[http://dx.doi.org/10.1111/j.1365-313X.2004.02002.x] [PMID: 14871316]

[85]   Shah SJ, Anjam MS, Mendy B, *et al.* Damage-associated responses of the host contribute to defence against cyst nematodes but not root-knot nematodes. J Exp Bot 2017; 68(21-22): 5949-60.
[http://dx.doi.org/10.1093/jxb/erx374] [PMID: 29053864]

[86]   Teixeira MA, Wei L, Kaloshian I. Root-knot nematodes induce pattern-triggered immunity in *Arabidopsis thaliana* roots. New Phytol 2016; 211(1): 276-87.
[http://dx.doi.org/10.1111/nph.13893] [PMID: 26892116]

[87]   Booker MA, DeLong A. Producing the ethylene signal: Regulation and diversification of ethylene biosynthetic enzymes. Plant Physiol 2015; 169(1): 42-50.
[http://dx.doi.org/10.1104/pp.15.00672] [PMID: 26134162]

[88]   Guan R, Su J, Meng X, *et al.* Multilayered regulation of ethylene induction plays a positive role in *Arabidopsis* resistance against *Pseudomonas syringae.* Plant Physiol 2015; 169(1): 299-312.
[http://dx.doi.org/10.1104/pp.15.00659] [PMID: 26265775]

[89]   Marhavý P, Kurenda A, Siddique S, *et al.* Single-cell damage elicits regional, nematode-restricting ethylene responses in roots. EMBO J 2019; 38(10): e100972.
[http://dx.doi.org/10.15252/embj.2018100972] [PMID: 31061171]

[90]   Fudali SL, Wang C, Williamson VM. Ethylene signaling pathway modulates attractiveness of host roots to the root-knot nematode *Meloidogyne hapla.* Mol Plant Microbe Interact 2013; 26(1): 75-86.
[http://dx.doi.org/10.1094/MPMI-05-12-0107-R] [PMID: 22712507]

[91]   Hu Y, You J, Li C, Williamson VM, Wang C. Ethylene response pathway modulates attractiveness of plant roots to soybean cyst nematode *Heterodera glycines.* Sci Rep 2017; 7(1): 41282.
[http://dx.doi.org/10.1038/srep41282] [PMID: 28112257]

[92]   Lorenzo O, Chico JM, Saénchez-Serrano JJ, Solano R. JASMONATE-INSENSITIVE1 encodes a MYC transcription factor essential to discriminate between different jasmonate-regulated defense responses in *Arabidopsis.* Plant Cell 2004; 16(7): 1938-50.
[http://dx.doi.org/10.1105/tpc.022319] [PMID: 15208388]

[93]   Pieterse CMJ, van Wees SCM, van Pelt JA, *et al.* A novel signaling pathway controlling induced systemic resistance in *Arabidopsis.* Plant Cell 1998; 10(9): 1571-80.
[http://dx.doi.org/10.1105/tpc.10.9.1571] [PMID: 9724702]

[94]   Zhang F, Yao J, Ke J, *et al.* Structural basis of JAZ repression of MYC transcription factors in jasmonate signalling. Nature 2015; 525(7568): 269-73.
[http://dx.doi.org/10.1038/nature14661] [PMID: 26258305]

[95]   Zhu Z, An F, Feng Y, *et al.* Derepression of ethylene-stabilized transcription factors (EIN_3/EIL_1) mediates jasmonate and ethylene signaling synergy in *Arabidopsis.* Proc Natl Acad Sci USA 2011; 108(30): 12539-44.
[http://dx.doi.org/10.1073/pnas.1103959108] [PMID: 21737749]

[96]   Glazebrook J. Contrasting mechanisms of defense against biotrophic and necrotrophic pathogens. Annu Rev Phytopathol 2005; 43(1): 205-27.
[http://dx.doi.org/10.1146/annurev.phyto.43.040204.135923] [PMID: 16078883]

[97]   Howe GA, Jander G. Plant immunity to insect herbivores. Annu Rev Plant Biol 2008; 59(1): 41-66.
[http://dx.doi.org/10.1146/annurev.arplant.59.032607.092825] [PMID: 18031220]

[98]   Nahar K, Kyndt T, De Vleesschauwer D, Höfte M, Gheysen G. The jasmonate pathway is a key player in systemically induced defense against root knot nematodes in rice. Plant Physiol 2011; 157(1): 305-

16.
[http://dx.doi.org/10.1104/pp.111.177576] [PMID: 21715672]

[99]   Thines B, Katsir L, Melotto M, *et al.* JAZ repressor proteins are targets of the SCF(COI1) complex during jasmonate signalling. Nature 2007; 448(7154): 661-5.
[http://dx.doi.org/10.1038/nature05960] [PMID: 17637677]

[100] Gupta A, Hisano H, Hojo Y, *et al.* Global profiling of phytohormone dynamics during combined drought and pathogen stress in *Arabidopsis thaliana* reveals ABA and JA as major regulators. Sci Rep 2017; 7(1): 4017.
[http://dx.doi.org/10.1038/s41598-017-03907-2] [PMID: 28638069]

[101] Ruan J, Zhou Y, Zhou M, *et al.* Jasmonic acid signaling pathway in plants. Int J Mol Sci 2019; 20(10): 2479.
[http://dx.doi.org/10.3390/ijms20102479] [PMID: 31137463]

[102] Khanam S, Bauters L, Singh RR, *et al.* Mechanisms of resistance in the rice cultivar Manikpukha to the rice stem nematode *Ditylenchus angustus*. Mol Plant Pathol 2018; 19(6): 1391-402.
[http://dx.doi.org/10.1111/mpp.12622] [PMID: 28990717]

[103] Kyndt T, Nahar K, Haegeman A, De Vleesschauwer D, Höfte M, Gheysen G. Comparing systemic defence-related gene expression changes upon migratory and sedentary nematode attack in rice. Plant Biol 2012; 14(s1) (Suppl. 1): 73-82.
[http://dx.doi.org/10.1111/j.1438-8677.2011.00524.x] [PMID: 22188265]

[104] Nahar K, Kyndt T, Nzogela YB, Gheysen G. Abscisic acid interacts antagonistically with classical defense pathways in rice–migratory nematode interaction. New Phytol 2012; 196(3): 901-13.
[http://dx.doi.org/10.1111/j.1469-8137.2012.04310.x] [PMID: 22985247]

[105] Bari R, Jones JDG. Role of plant hormones in plant defence responses. Plant Mol Biol 2009; 69(4): 473-88.
[http://dx.doi.org/10.1007/s11103-008-9435-0] [PMID: 19083153]

[106] López MA, Bannenberg G, Castresana C. Controlling hormone signaling is a plant and pathogen challenge for growth and survival. Curr Opin Plant Biol 2008; 11(4): 420-7.
[http://dx.doi.org/10.1016/j.pbi.2008.05.002] [PMID: 18585953]

[107] Dempsey DMA, Vlot AC, Wildermuth MC, Klessig DF. Salicylic Acid biosynthesis and metabolism. *Arabidopsis* Book 2011; 9: e0156.
[http://dx.doi.org/10.1199/tab.0156] [PMID: 22303280]

[108] Durrant WE, Wang S, Dong X. *Arabidopsis* SNI1 and RAD51D regulate both gene transcription and DNA recombination during the defense response. Proc Natl Acad Sci USA 2007; 104(10): 4223-7.
[http://dx.doi.org/10.1073/pnas.0609357104] [PMID: 17360504]

[109] Wilson SK, Pretorius T, Naidoo S. Mechanisms of systemic resistance to pathogen infection in plants and their potential application in forestry. BMC Plant Biology. 2023 Aug 25; 23(1): 404.
[http://dx.doi.org/10.1186/s12870-023-04391-9]

[110] Pieterse CMJ, Van der Does D, Zamioudis C, Leon-Reyes A, Van Wees SCM. Hormonal modulation of plant immunity. Annu Rev Cell Dev Biol 2012; 28(1): 489-521.
[http://dx.doi.org/10.1146/annurev-cellbio-092910-154055] [PMID: 22559264]

[111] Chisholm ST, Coaker G, Day B, Staskawicz BJ. Host-microbe interactions: Shaping the evolution of the plant immune response. Cell 2006; 124(4): 803-14.
[http://dx.doi.org/10.1016/j.cell.2006.02.008] [PMID: 16497589]

[112] Jones JDG, Dangl JL. The plant immune system. Nature 2006; 444(7117): 323-9.
[http://dx.doi.org/10.1038/nature05286] [PMID: 17108957]

[113] Dodds PN, Rathjen JP. Plant immunity: Towards an integrated view of plant–pathogen interactions. Nat Rev Genet 2010; 11(8): 539-48.
[http://dx.doi.org/10.1038/nrg2812] [PMID: 20585331]

[114] Van der Biezen EA, Jones JD. Plant disease-resistance proteins and the gene-for-gene concept. Trends Biochem Sci 1998; 23(12): 454-6.
[http://dx.doi.org/10.1016/S0968-0004(98)01311-5] [PMID: 9868361]

[115] Navarro L, Zipfel C, Rowland O, *et al.* The transcriptional innate immune response to flg22. Interplay and overlap with Avr gene-dependent defense responses and bacterial pathogenesis. Plant Physiol 2004; 135(2): 1113-28.
[http://dx.doi.org/10.1104/pp.103.036749] [PMID: 15181213]

[116] Nguyen QM, Iswanto ABB, Son GH, Kim SH. Recent Advances in effector-triggered immunity in plants: New Pieces in the puzzle create a different paradigm. Int J Mol Sci 2021; 22(9): 4709.
[http://dx.doi.org/10.3390/ijms22094709] [PMID: 33946790]

[117] Tao Y, Xie Z, Chen W, *et al.* Quantitative nature of *Arabidopsis* responses during compatible and incompatible interactions with the bacterial pathogen *Pseudomonas syringae.* Plant Cell 2003; 15(2): 317-30.
[http://dx.doi.org/10.1105/tpc.007591] [PMID: 12566575]

[118] Richard MMS, Gratias A, Meyers BC, Geffroy V. Molecular mechanisms that limit the costs of NLR-mediated resistance in plants. Mol Plant Pathol 2018; 19(11): 2516-23.
[http://dx.doi.org/10.1111/mpp.12723] [PMID: 30011120]

[119] Yuan M, Jiang Z, Bi G, *et al.* Pattern-recognition receptors are required for NLR-mediated plant immunity. Nature 2021; 592(7852): 105-9.
[http://dx.doi.org/10.1038/s41586-021-03316-6] [PMID: 33692546]

[120] Gao Y, Wu Y, Du J, *et al.* Both light-induced SA accumulation and ETI mediators contribute to the cell death regulated by BAK1 and BKK1. Front Plant Sci 2017; 8: 622.
[http://dx.doi.org/10.3389/fpls.2017.00622] [PMID: 28487714]

[121] Ngou BPM, Ahn HK, Ding P, Jones JDG. Mutual potentiation of plant immunity by cell-surface and intracellular receptors. Nature 2021; 592(7852): 110-5.
[http://dx.doi.org/10.1038/s41586-021-03315-7] [PMID: 33692545]

[122] van Wersch S, Li X. Stronger When Together: Clustering of plant NLR disease resistance genes. Trends Plant Sci 2019; 24(8): 688-99.
[http://dx.doi.org/10.1016/j.tplants.2019.05.005] [PMID: 31266697]

[123] Williamson MV, Kumar A. Nematode resistance in plants: The battle underground. Trends Genet 2006; 22(7): 396-403.
[http://dx.doi.org/10.1016/j.tig.2006.05.003] [PMID: 16723170]

[124] Liu S, Kandoth PK, Warren SD, *et al.* A soybean cyst nematode resistance gene points to a new mechanism of plant resistance to pathogens. Nature 2012; 492(7428): 256-60.
[http://dx.doi.org/10.1038/nature11651] [PMID: 23235880]

[125] Liu S, Kandoth PK, Lakhssassi N, *et al.* The soybean GmSNAP18 gene underlies two types of resistance to soybean cyst nematode. Nat Commun 2017; 8(1): 14822.
[http://dx.doi.org/10.1038/ncomms14822] [PMID: 28345654]

[126] Peng Y, van Wersch R, Zhang Y. Convergent and divergent signaling in PAMP-triggered immunity and effector-triggered immunity. Mol Plant Microbe Interact 2018; 31(4): 403-9.
[http://dx.doi.org/10.1094/MPMI-06-17-0145-CR] [PMID: 29135338]

[127] Cai D, Kleine M, Kifle S, *et al.* Positional cloning of a gene for nematode resistance in sugar beet. Science 1997; 275(5301): 832-4.
[http://dx.doi.org/10.1126/science.275.5301.832] [PMID: 9012350]

[128] Ernst K, Kumar A, Kriseleit D, Kloos DU, Phillips MS, Ganal MW. The broad-spectrum potato cyst nematode resistance gene (Hero) from tomato is the only member of a large gene family of NBS-LRR genes with an unusual amino acid repeat in the LRR region. Plant J 2002; 31(2): 127-36.
[http://dx.doi.org/10.1046/j.1365-313X.2002.01341.x] [PMID: 12121443]

[129] McLean MD, Hoover GJ, Bancroft B, *et al.* Identification of the full-length *Hs1* pro-l coding sequence and preliminary evaluation of soybean cyst nematode resistance in soybean transformed with *Hs1* pro-l cDNA. Can J Bot 2007; 85(4): 437-41.
[http://dx.doi.org/10.1139/B07-038]

[130] Milligan SB, Bodeau J, Yaghoobi J, Kaloshian I, Zabel P, Williamson VM. The root knot nematode resistance gene Mi from tomato is a member of the leucine zipper, nucleotide binding, leucine-rich repeat family of plant genes. Plant Cell 1998; 10(8): 1307-19.
[http://dx.doi.org/10.1105/tpc.10.8.1307] [PMID: 9707531]

[131] Thurau T, Kifle S, Jung C, Cai D. The promoter of the nematode resistance gene Hs1pro-1 activates a nematode-responsive and feeding site-specific gene expression in sugar beet (*Beta vulgaris* L.) and *Arabidopsis thaliana*. Plant Mol Biol 2003; 52(3): 643-60.
[http://dx.doi.org/10.1023/A:1024887516581] [PMID: 12956533]

[132] Van der Vossen EAG, Van der Voort JN, Kanyuka K, *et al.* Homologues of a single resistance-gene cluster in potato confer resistance to distinct pathogens: A virus and a nematode. Plant J 2000; 23(5): 567-76.
[http://dx.doi.org/10.1046/j.1365-313x.2000.00814.x] [PMID: 10972883]

[133] Bayless AM, Zapotocny RW, Han S, Grunwald DJ, Amundson KK, Bent AF. The *rhg1-a* (*Rhg1* low-copy) nematode resistance source harbors a copia-family retrotransposon within the *Rhg1*- encoded α-SNAP gene. Plant Direct 2019; 3(8): e00164.
[http://dx.doi.org/10.1002/pld3.164] [PMID: 31468029]

[134] Gillet FX, Bournaud C, de Souza Júnior JDA, Fatima Grossi-de-Sa M. Plant-parasitic nematodes: Towards understanding molecular players in stress responses. Ann Bot (Lond) 2017; 119(5): mcw260.
[http://dx.doi.org/10.1093/aob/mcw260] [PMID: 28087659]

[135] Ye D, Jiang Y, Wang C, Roberts PA. Expression analysis of MicroRNAs and their target genes in *Cucumis metuliferus* infected by the root-knot nematode *Meloidogyne incognita*. Physiol Mol Plant Pathol 2020; 111: 101491.
[http://dx.doi.org/10.1016/j.pmpp.2020.101491]

[136] Okamura K, Hagen JW, Duan H, Tyler DM, Lai EC. The mirtron pathway generates microRNA-class regulatory RNAs in *Drosophila*. Cell 2007; 130(1): 89-100.
[http://dx.doi.org/10.1016/j.cell.2007.06.028] [PMID: 17599402]

[137] Okamura K, Ladewig E, Zhou L, Lai EC. Functional small RNAs are generated from select miRNA hairpin loops in flies and mammals. Genes Dev 2013; 27(7): 778-92.
[http://dx.doi.org/10.1101/gad.211698.112] [PMID: 23535236]

[138] Yang JS, Lai EC. Alternative miRNA biogenesis pathways and the interpretation of core miRNA pathway mutants. Mol Cell 2011; 43(6): 892-903.
[http://dx.doi.org/10.1016/j.molcel.2011.07.024] [PMID: 21925378]

[139] Vijayapalani P, Hewezi T, Pontvianne F, Baum TJ. An effector from the cyst nematode *Heterodera schachtii* derepresses host rRNA genes by altering histone acetylation. Plant Cell 2018; 30(11): 2795-812.
[http://dx.doi.org/10.1105/tpc.18.00570] [PMID: 30333146]

[140] Cabrera J, Barcala M, García A, *et al.* Differentially expressed small RNAs in *Arabidopsis* galls formed by *Meloidogyne javanica* : A functional role for miR390 and its TAS 3-derived tasiRNAs. New Phytol 2016; 209(4): 1625-40.
[http://dx.doi.org/10.1111/nph.13735] [PMID: 26542733]

[141] Lei P, Qi N, Zhou Y, *et al.* Soybean miR159-GmMYB33 regulatory network involved in gibberellin-modulated resistance to *Heterodera glycines*. Int J Mol Sci 2021; 22(23): 13172.
[http://dx.doi.org/10.3390/ijms222313172] [PMID: 34884977]

[142] Boerma HR, Hussey RS. Breeding plants for resistance to nematodes. J Nematol 1992; 24(2): 242-52.

[PMID: 19282990]

[143] Fuller VL, Lilley CJ, Urwin PE. Nematode resistance. New Phytol 2008; 180(1): 27-44.
[http://dx.doi.org/10.1111/j.1469-8137.2008.02508.x] [PMID: 18564304]

[144] Lilley CJ, Devlin P, Urwin PE, Atkinson HJ. Parasitic nematodes, proteinases and transgenic plants. Trends Parasitol 1999. 15(10); 414-417.
[http://dx.doi.org/10.1016/S0169-4758(99)01513-6]

[145] Klink VP, Matthews BF. Emerging approaches to broaden resistance of soybean to soybean cyst nematode as supported by gene expression studies. Plant Physiol 2009; 151(3): 1017-22.
[http://dx.doi.org/10.1104/pp.109.144006] [PMID: 19675146]

[146] Siddique S, Endres S, Atkins JM, *et al.* Myo-inositol oxygenase genes are involved in the development of syncytia induced by *Heterodera schachtii* in *Arabidopsis* roots. New Phytol 2009; 184(2): 457-72.
[http://dx.doi.org/10.1111/j.1469-8137.2009.02981.x] [PMID: 19691674]

[147] Ali MA, Wieczorek K, Kreil DP, Bohlmann H. The beet cyst nematode *Heterodera schachtii* modulates the expression of WRKY transcription factors in syncytia to favour its development in *Arabidopsis* roots. PLoS One 2014; 9(7): e102360.
[http://dx.doi.org/10.1371/journal.pone.0102360] [PMID: 25033038]

[148] Atkinson HJ, Lilley CJ, Urwin PE. Strategies for transgenic nematode control in developed and developing world crops. Curr Opin Biotechnol 2012; 23(2): 251-6.
[http://dx.doi.org/10.1016/j.copbio.2011.09.004] [PMID: 21996368]

[149] Green J, Wang D, Lilley CJ, Urwin PE, Atkinson HJ. Transgenic potatoes for potato cyst nematode control can replace pesticide use without impact on soil quality. PLoS One 2012; 7(2): e30973.
[http://dx.doi.org/10.1371/journal.pone.0030973] [PMID: 22359559]

[150] Lilley CJ, Wang D, Atkinson HJ, Urwin PE. Effective delivery of a nematode-repellent peptide using a root-cap-specific promoter. Plant Biotechnol J 2011; 9(2): 151-61.
[http://dx.doi.org/10.1111/j.1467-7652.2010.00542.x] [PMID: 20602721]

[151] Wieczorek K, Hofmann J, Blöchl A, Szakasits D, Bohlmann H, Grundler FMW. *Arabidopsis* endo-1,4-β-glucanases are involved in the formation of root syncytia induced by *Heterodera schachtii*. Plant J 2008; 53(2): 336-51.
[http://dx.doi.org/10.1111/j.1365-313X.2007.03340.x] [PMID: 18069944]

[152] Amjad Ali M, Abbas A, Azeem F, Javed N, Bohlmann H. Plant-nematode interactions: From genomics to metabolomics. Int J Agric Biol 2015; 17(6): 1071-82.
[http://dx.doi.org/10.17957/IJAB/15.0037]

[153] Barcala M, García A, Cabrera J, *et al.* Early transcriptomic events in microdissected *Arabidopsis* nematode-induced giant cells. Plant J 2010; 61(4): 698-712.
[http://dx.doi.org/10.1111/j.1365-313X.2009.04098.x] [PMID: 20003167]

[154] Szakasits D, Heinen P, Wieczorek K, *et al.* The transcriptome of syncytia induced by the cyst nematode *Heterodera schachtii* in *Arabidopsis* roots. Plant J 2009; 57(5): 771-84.
[http://dx.doi.org/10.1111/j.1365-313X.2008.03727.x] [PMID: 18980640]

[155] Birkenbihl RP, Diezel C, Somssich IE. *Arabidopsis* WRKY33 is a key transcriptional regulator of hormonal and metabolic responses toward *Botrytis cinerea* infection. Plant Physiol 2012; 159(1): 266-85.
[http://dx.doi.org/10.1104/pp.111.192641] [PMID: 22392279]

[156] Hofmann J, Youssef-Banora M, de Almeida-Engler J, Grundler FMW. The role of callose deposition along plasmodesmata in nematode feeding sites. Mol Plant Microbe Interact 2010; 23(5): 549-57.
[http://dx.doi.org/10.1094/MPMI-23-5-0549] [PMID: 20367463]

[157] Mao G, Meng X, Liu Y, Zheng Z, Chen Z, Zhang S. Phosphorylation of a WRKY transcription factor by two pathogen-responsive MAPKs drives phytoalexin biosynthesis in *Arabidopsis*. Plant Cell 2011; 23(4): 1639-53.

[http://dx.doi.org/10.1105/tpc.111.084996] [PMID: 21498677]

[158] Millet YA, Danna CH, Clay NK, *et al.* Innate immune responses activated in *Arabidopsis* roots by microbe-associated molecular patterns. Plant Cell 2010; 22(3): 973-90.
[http://dx.doi.org/10.1105/tpc.109.069658] [PMID: 20348432]

[159] Youssef RM, MacDonald MH, Brewer EP, Bauchan GR, Kim KH, Matthews BF. Ectopic expression of AtPAD4 broadens resistance of soybean to soybean cyst and root-knot nematodes. BMC Plant Biol 2013; 13(1): 67.
[http://dx.doi.org/10.1186/1471-2229-13-67] [PMID: 23617694]

[160] Lin J, Mazarei M, Zhao N, *et al.* Overexpression of a soybean salicylic acid methyltransferase gene confers resistance to soybean cyst nematode. Plant Biotechnol J 2013; 11(9): 1135-45.
[http://dx.doi.org/10.1111/pbi.12108] [PMID: 24034273]

[161] Sorribas FJ, Djian-Caporalino C, Mateille T. Nematodes In: Gullino ML, Albajes R, Nicot PC, Eds. Integrated pest and disease management in greenhouse crops Cham: Springer International Publishing 2020; pp. 147-74.
[http://dx.doi.org/10.1007/978-3-030-22304-5_5]

[162] Khan MR. Nematode pests of agricultural crops, a global overview. In: Khan MR, Ed. Novel biological and biotechnological applications in plant nematode management Singapore: Springer. Nat Singap 2023; 3-45.
[http://dx.doi.org/10.1007/978-981-99-2893-4_1]

[163] Anwar SA, Zia A, Javed N, *et al.* Weeds as reservoir of nematodes. Pak J Nematol 2009; 27: 145-53.

[164] Rich JR. Brito JA, Kaur R, Ferrell JA. Weed species as hosts of *Meloidogyne*: A review. Nematropica 2009; 39: 157-85.

[165] Lawrence KS, Price AJ, Lawrence GW, Jones JR, Akridge JR. Weed hosts for *Rotylenchulus reniformis* in cotton fields rotated with corn in the Southeast of the United States (USA), the reniform nematode (*Rotylenchulus reniformis*) Linford & Oliveira is the primary nematode pest of cotton. Nematropica 2008; 38: 13-22.

[166] Mateille T, Schwey D, Amazouz S. Au Maroc, un outil de gestion des nématodes *Meloidogyne* spp. en production maraîchère. Phytoma - Déf Végétaux. 2005; 84: 40–43.

[167] McSorley R. Alternative practices for managing plant-parasitic nematodes. Am J Altern Agric 1998; 13(3): 98-104.
[http://dx.doi.org/10.1017/S0889189300007761]

[168] Zasada IA, Halbrendt JM, Kokalis-Burelle N, LaMondia J, McKenry MV, Noling JW. Managing nematodes without methyl bromide. Annu Rev Phytopathol 2010; 48(1): 311-28.
[http://dx.doi.org/10.1146/annurev-phyto-073009-114425] [PMID: 20455696]

[169] Timper P. Conserving and enhancing biological control of nematodes. J Nematol 2014; 46(2): 75-89.
[PMID: 24987159]

[170] Abawi GS, Widmer TL. Impact of soil health management practices on soilborne pathogens, nematodes and root diseases of vegetable crops. Appl Soil Ecol 2000; 15(1): 37-47.
[http://dx.doi.org/10.1016/S0929-1393(00)00070-6]

[171] Ball BC, Bingham I, Rees RM, Watson CA, Litterick A. The role of crop rotations in determining soil structure and crop growth conditions. Can J Soil Sci 2005; 85(5): 557-77.
[http://dx.doi.org/10.4141/S04-078]

[172] Page K, Dang Y, Dalal R. Impacts of conservation tillage on soil quality, including soil-borne crop diseases, with a focus on semi-arid grain cropping systems. Australas Plant Pathol 2013; 42(3): 363-77.
[http://dx.doi.org/10.1007/s13313-013-0198-y]

[173] Reeve JR, Hoagland LA, Villalba JJ, *et al.* Organic farming, soil health, and food quality: Considering

possible links. In: Advances in Agronomy. 2016; 137: 2016, 319-367.
[http://dx.doi.org/10.1016/bs.agron.2015.12.003]

[174]   Norris CE, Congreves KA. Alternative management practices improve soil health indices in intensive vegetable cropping systems: A review. Front Environ Sci 2018; 6: 50.
[http://dx.doi.org/10.3389/fenvs.2018.00050]

[175]   Shah KK, Modi B, Pandey HP, *et al.* Diversified crop rotation: An approach for sustainable agriculture production Adv Agric. 2021; pp. 1-9.
[http://dx.doi.org/10.1155/2021/8924087]

[176]   Lien G, Brian Hardaker J, Flaten O. Risk and economic sustainability of crop farming systems. Agric Syst 2007; 94(2): 541-52.
[http://dx.doi.org/10.1016/j.agsy.2007.01.006]

[177]   Starr JL, Koenning SR, Kirkpatrick TL, Robinson AF, Roberts PA, Nichols RL. The future of nematode management in cotton. J Nematol 2007; 39(4): 283-94.
[PMID: 19259500]

[178]   Hayati D, Ranjbar Z, Karami E. Measuring agricultural sustainability. In: Lichtfouse E Ed. Biodiversity, biofuels, agroforestry and conservation agriculture. 2010; pp. 73-100.

[179]   Reganold JP, Wachter JM. Organic agriculture in the twenty-first century. Nat Plants 2016; 2(2): 15221.
[http://dx.doi.org/10.1038/nplants.2015.221] [PMID: 27249193]

[180]   Yu T, Mahe L, Li Y, Wei X, Deng X, Zhang D. Benefits of crop rotation on climate resilience and its prospects in China. Agronomy (Basel) 2022; 12(2): 436.
[http://dx.doi.org/10.3390/agronomy12020436]

[181]   Hooks CRR, Wang KH, Ploeg A, McSorley R. Using marigold (*Tagetes* spp.) as a cover crop to protect crops from plant-parasitic nematodes. Appl Soil Ecol 2010; 46(3): 307-20.
[http://dx.doi.org/10.1016/j.apsoil.2010.09.005]

[182]   Dutta TK, Khan MR, Phani V. Plant-parasitic nematode management *via* biofumigation using brassica and non-brassica plants: Current status and future prospects. Curr Plant Biol 2019; 17: 17-32.
[http://dx.doi.org/10.1016/j.cpb.2019.02.001]

[183]   Westphal A. Sustainable approaches to the management of plant-parasitic nematodes and disease complexes. J Nematol 2011; 43(2): 122-5.
[PMID: 22791923]

[184]   Widmer TL, Mitkowski NA, Abawi GS. Soil organic matter and management of plant-parasitic nematodes. J Nematol 2002; 34(4): 289-95.
[PMID: 19265946]

[185]   Daryanto S, Fu B, Wang L, Jacinthe PA, Zhao W. Quantitative synthesis on the ecosystem services of cover crops. Earth Sci Rev 2018; 185: 357-73.
[http://dx.doi.org/10.1016/j.earscirev.2018.06.013]

[186]   Tian Y, Zhang X, Liu J, Gao L. Effects of summer cover crop and residue management on cucumber growth in intensive Chinese production systems: Soil nutrients, microbial properties and nematodes. Plant Soil 2011; 339(1-2): 299-315.
[http://dx.doi.org/10.1007/s11104-010-0579-8]

[187]   Djigal D, Chabrier C, Duyck PF, Achard R, Quénéhervé P, Tixier P. Cover crops alter the soil nematode food web in banana agroecosystems. Soil Biol Biochem 2012; 48: 142-50.
[http://dx.doi.org/10.1016/j.soilbio.2012.01.026]

[188]   Ralmi NHAA, Khandaker MM, Mat N. Occurrence and control of root knot nematode in crops: A review. Aust J Crop Sci 2016; 10(12): 1649-54.
[http://dx.doi.org/10.21475/ajcs.2016.10.12.p7444]

[189] Abd-Elgawad MMM. Optimizing Safe approaches to manage plant-parasitic nematodes. plants. 2021; 10: 1911.
[http://dx.doi.org/10.3390/plants10091911]

[190] Fourie H, Ahuja P, Lammers J, Daneel M. Brassicacea-based management strategies as an alternative to combat nematode pests: A synopsis. Crop Prot 2016; 80: 21-41.
[http://dx.doi.org/10.1016/j.cropro.2015.10.026]

[191] Westphal A, Scott AW. Implementation of soybean in cotton cropping sequences for management of reniform nematode in South Texas. Crop Sci. 2005; 45: cropsci2005.0233.
[http://dx.doi.org/10.2135/cropsci2005.0233]

[192] Abebe E, Mekete T, Seid A, *et al.* Research on plant-parasitic and entomopathogenic nematodes in Ethiopia: A review of current state and future direction. Nematology 2015; 17(7): 741-59.
[http://dx.doi.org/10.1163/15685411-00002919]

[193] Shiferaw T, Dechassa N, Sakhuja PK. Influence of poultry litter and rapeseed cake on infestation of *Meloidogyne incognita* on tomato in Dire Dawa, Eastern Ethiopia. 2014; 32: 67–72.

[194] Rao MS, Umamaheswari R, Chakravarthy AK, Manojkumar R, Rajinikanth R, Chaya MK, K. P. Nematode Management in corn nematode management in corn. IIHR Tech Bull No 48 ICAR- Indian Inst Hortic Res Bengaluru India, 2015.
[http://dx.doi.org/10.1007/978-981-15-0794-6_24]

[195] Rosskopf E, Di Gioia F, Hong JC, Pisani C, Kokalis-Burelle N. Organic amendments for pathogen and nematode control. Annu Rev Phytopathol 2020; 58(1): 277-311.
[http://dx.doi.org/10.1146/annurev-phyto-080516-035608] [PMID: 32853099]

[196] McSorley R. Overview of organic amendments for management of plant-parasitic nematodes, with case studies from Florida. J Nematol 2011; 43(2): 69-81.
[PMID: 22791915]

[197] Giné A, Sorribas FJ. Effect of plant resistance and BioAct WG (*Purpureocillium lilacinum* strain 251) on *Meloidogyne incognita* in a tomato–cucumber rotation in a greenhouse. Pest Manag Sci 2017; 73(5): 880-7.
[http://dx.doi.org/10.1002/ps.4357] [PMID: 27414251]

[198] Desaeger J, Wram C, Zasada I. New reduced-risk agricultural nematicides - rationale and review. J Nematol 2020; 52(1): 1-16.
[http://dx.doi.org/10.21307/jofnem-2020-091] [PMID: 33829179]

[199] Oka Y. From old-generation to next-generation nematicides. Agronomy (Basel) 2020; 10(9): 1387.
[http://dx.doi.org/10.3390/agronomy10091387]

[200] Giannakou IO, Sidiropoulos A, Prophetou-Athanasiadou D. Chemical alternatives to methyl bromide for the control of root-knot nematodes in greenhouses. Pest Manag Sci 2002; 58(3): 290-6.
[http://dx.doi.org/10.1002/ps.453] [PMID: 11975176]

[201] Oloo G, Aguyoh JN. O. TG, Ombiri OJ. Alternative management strategies for weeds and root knot nematodes (*Meloidogyne* spp) in rose plants grown under polyethylene covered tunnels. ARPN J Agric Biol Sci 2009; 4: 23-8.

[202] Desaeger J, Csinos A, Timper P, Hammes G, Seebold K. Soil fumigation and oxamyl drip applications for nematode and insect control in vegetable plasticulture. Ann Appl Biol 2004; 145(1): 59-70.
[http://dx.doi.org/10.1111/j.1744-7348.2004.tb00359.x]

[203] Kearn J, Ludlow E, Dillon J, O'Connor V, Holden-Dye L. Fluensulfone is a nematicide with a mode of action distinct from anticholinesterases and macrocyclic lactones. Pestic Biochem Physiol 2014; 109: 44-57.
[http://dx.doi.org/10.1016/j.pestbp.2014.01.004] [PMID: 24581383]

[204] Lahm GP, Desaeger J, Smith BK, *et al.* The discovery of fluazaindolizine: A new product for the

control of plant parasitic nematodes. Bioorg Med Chem Lett 2017; 27(7): 1572-5.
[http://dx.doi.org/10.1016/j.bmcl.2017.02.029] [PMID: 28242274]

[205]   Chawla S, Patel DJ, Patel SH, Kalasariya RL, Shah PG. Behaviour and risk assessment of fluopyram and its metabolite in cucumber (*Cucumis sativus*) fruit and in soil. Environ Sci Pollut Res Int 2018; 25(12): 11626-34.
[http://dx.doi.org/10.1007/s11356-018-1439-y] [PMID: 29429109]

[206]   Giannakou IO, Panopoulou S. The use of fluensulfone for the control of root-knot nematodes in greenhouse cultivated crops: Efficacy and phytotoxicity effects. Cogent Food Agric. 2019; 5: 1643819.
[http://dx.doi.org/10.1080/23311932.2019.1643819]

[207]   Oka Y, Saroya Y. Effect of fluensulfone and fluopyram on the mobility and infection of second-stage juveniles of *Meloidogyne incognita* and *M. javanica*. Pest Manag Sci 2019; 75(8): 2095-106.
[http://dx.doi.org/10.1002/ps.5399] [PMID: 30843368]

[208]   Thoden TC, Wiles JA. Biological attributes of Salibro™, a novel sulfonamide nematicide. Part 1: Impact on the fitness of *Meloidogyne incognita, M. hapla* and *Acrobeloides buetschlii*. Nematology 2019; 21(6): 625-39.
[http://dx.doi.org/10.1163/15685411-00003240]

[209]   Schleker ASS, Rist M, Matera C, *et al.* Mode of action of fluopyram in plant-parasitic nematodes. Sci Rep 2022; 12(1): 11954.
[http://dx.doi.org/10.1038/s41598-022-15782-7] [PMID: 35831379]

[210]   Burns AR, Luciani GM, Musso G, *et al. Caenorhabditis elegans* is a useful model for anthelmintic discovery. Nat Commun 2015; 6(1): 7485.
[http://dx.doi.org/10.1038/ncomms8485] [PMID: 26108372]

[211]   Sande D, Mullen J, Wetzstein M, Houston J. Environmental impacts from pesticide use: A case study of soil fumigation in Florida tomato production. Int J Environ Res Public Health 2011; 8(12): 4649-61.
[http://dx.doi.org/10.3390/ijerph8124649] [PMID: 22408594]

[212]   Noling JW. Nematode management in tomatoes, peppers and eggplant. Univ Fla IFAS Ext 2019; ENY-032: 1-15.

[213]   Liu C, Grabau Z. *Meloidogyne incognita* management using fumigant and non-fumigant nematicides on sweet potato. J Nematol 2022; 54(1): 20220026.
[http://dx.doi.org/10.2478/jofnem-2022-0026] [PMID: 35975221]

[214]   Alam MS, Khanal C, Roberts J, Rutter W. Impact of non-fumigant nematicides on reproduction and pathogenicity of *Meloidogyne enterolobii* and disease severity in tobacco. J Nematol 2023; 55(1): 20230025.
[http://dx.doi.org/10.2478/jofnem-2023-0025] [PMID: 37284001]

[215]   Poveda J, Abril-Urias P, Escobar C. Biological Control of plant-parasitic nematodes by filamentous fungi inducers of resistance: *Trichoderma*, mycorrhizal and endophytic fungi. Front Microbiol 2020; 11: 992.
[http://dx.doi.org/10.3389/fmicb.2020.00992] [PMID: 32523567]

[216]   Ahmad G, Khan A, Khan AA, Ali A, Mohhamad HI. Biological control: A novel strategy for the control of the plant parasitic nematodes. Antonie van Leeuwenhoek 2021; 114(7): 885-912.
[http://dx.doi.org/10.1007/s10482-021-01577-9] [PMID: 33893903]

[217]   Sathyan T, Dhanya MK, Murugan M, *et al.* Evaluation of bio-agents, synthetic insecticides and organic amendment against the root-knot nematode, *Meloidogyne* spp in cardamom. J Biol Control 2021; 35: 68-75. [*Elettaria cardamomum* (L.) Maton].
[http://dx.doi.org/10.18311/jbc/2021/26772]

[218]   Krif G, Lahlali R, El Aissami A, *et al.* Efficacy of authentic bio-nematicides against the root-knot nematode, *Meloidogyne javanica* infecting tomato under greenhouse conditions. Physiol Mol Plant

Pathol 2022; 118: 101803.
[http://dx.doi.org/10.1016/j.pmpp.2022.101803]

[219] Poveda J, Eugui D. Combined use of *Trichoderma* and beneficial bacteria (mainly *Bacillus* and *Pseudomonas*): Development of microbial synergistic bio-inoculants in sustainable agriculture. Biol Control 2022; 176: 105100.
[http://dx.doi.org/10.1016/j.biocontrol.2022.105100]

[220] Saad AM, Salem HM, El-Tahan AM, *et al.* Biological control: An effective approach against nematodes using black pepper plants (*Piper nigrum* L.). Saudi J Biol Sci 2022; 29(4): 2047-55.
[http://dx.doi.org/10.1016/j.sjbs.2022.01.004] [PMID: 35531173]

[221] Park JH, Kim SJ, Choi JH, Yoon MH, Chung DY, Kim HJ. Biological control by nematophagous fungi for plant-parasitic nematodes in soils. Korean J Soil Sci Fertil 2012; 45(1): 74-8.
[http://dx.doi.org/10.7745/KJSSF.2012.45.1.074]

[222] Dasgupta MK, Khan MR. Nematophagous fungi: Ecology, diversity and geographical distribution Biocontrol Agents Phytonematodes. UK: CABI 2015; pp. 126-62.
[http://dx.doi.org/10.1079/9781780643755.0126]

[223] Zhang Y, Li S, Li H, Wang R, Zhang KQ, Xu J. Fungi–nematode interactions: Diversity, ecology, and biocontrol prospects in agriculture. J Fungi (Basel) 2020; 6(4): 206.
[http://dx.doi.org/10.3390/jof6040206] [PMID: 33020457]

[224] Al-Ani LKT, Soares FEF, Sharma A, de los Santos-Villalobos S, Valdivia-Padilla AV, Aguilar-Marcelino L. Strategy of nematophagous fungi in determining the activity of plant parasitic nematodes and their prospective role in sustainable agriculture. Front Fungal Biol 2022; 3: 863198.
[http://dx.doi.org/10.3389/ffunb.2022.863198] [PMID: 37746161]

[225] Saeed M, Mukhtar T, Ahmed R, Ahmad T, Iqbal MA. Suppression of *Meloidogyne javanica* infection in peach (*Prunus persica* (L.) Batsch) using fungal biocontrol agents. Sustainability (Basel) 2023; 15(18): 13833.
[http://dx.doi.org/10.3390/su151813833]

[226] Bordallo JJ, Lopez-Llorca LV, Jansson HB, Salinas J, Persmark L, Asensio L. Colonization of plant roots by egg-parasitic and nematode-trapping fungi. New Phytol 2002; 154(2): 491-9.
[http://dx.doi.org/10.1046/j.1469-8137.2002.00399.x] [PMID: 33873431]

[227] Lopez-Llorca LV, Bordallo JJ, Salinas J, Monfort E, López-Serna ML. Use of light and scanning electron microscopy to examine colonisation of barley rhizosphere by the nematophagous fungus *Verticillium chlamydosporium*. Micron 2002; 33(1): 61-7.
[http://dx.doi.org/10.1016/S0968-4328(00)00070-6] [PMID: 11473815]

[228] Aguilar-Marcelino L, Mendoza-de-Gives P, Al-Ani LKT, *et al.* Using molecular techniques applied to beneficial microorganisms as biotechnological tools for controlling agricultural plant pathogens and pest In: Sharma V, Salwan R, Al-Ani LKT, Eds, Molecular aspects of plant beneficial microbes in agriculture. Elsevier 2020; pp. 333-49.
[http://dx.doi.org/10.1016/B978-0-12-818469-1.00027-4]

[229] Sri Hastuti LD, Faull J. Wheat bran soil inoculant of sumateran nematode-trapping fungi as biocontrol agents of the root-knot nematode *Meloidogyne incognita* on deli tobacco (*Nicotiana tabaccum* l) cv. deli 4. IOP Conf Ser Earth Environ Sci 2018; 130: 012009.
[http://dx.doi.org/10.1088/1755-1315/130/1/012009]

[230] Jansson HB, Jeyaprakash A, Zuckerman BM. Control of root-knot nematodes on tomato by the endoparasitic fungus *Meria coniospora*. J Nematol 1985; 17(3): 327-9.
[PMID: 19294101]

[231] Singh UB, Sahu A, Sahu N, *et al.* *Arthrobotrys oligospora* -mediated biological control of diseases of tomato (*Lycopersicon esculentum* Mill.) caused by *Meloidogyne incognita* and *Rhizoctonia solani*. J Appl Microbiol 2013; 114(1): 196-208.
[http://dx.doi.org/10.1111/jam.12009] [PMID: 22963133]

[232] de Gives PM, Braga FR. *Pochonia chlamydosporia*: A promising biotechnological tool against parasitic nematodes and geohelminths. In: Manzanilla-López RH, Lopez-Llorca LV Eds., Perspectives in sustainable nematode management through Pochonia chlamydosporia applications for root and rhizosphere health Cham: Springer International Publishing 2017; pp. 371-83.
[http://dx.doi.org/10.1007/978-3-319-59224-4_17]

[233] Devi G. Utilization of nematode destroying fungi for management of plant-parasitic nematodes-A review. Biosci Biotechnol Res Asia 2018; 15(2): 377-96.
[http://dx.doi.org/10.13005/bbra/2642]

[234] Siddiqui IA, Qureshi SA, Sultana V, Ehteshamul-Haque S, Ghaffar A. Biological control of root rot-root knot disease complex of tomato. Plant Soil 2000; 227(1/2): 163-9.
[http://dx.doi.org/10.1023/A:1026599532684]

[235] Cumagun CJR. Advances in formulation of *Trichoderma* for biocontrol. In: Gupta VK, Schmoll M, Herrera-Estrella A, Upadhyay RS, Druzhinina I, Tuohy MG, Eds., Biotechnology and biology of Trichoderma Elsevier 2014; pp. 527-31.
[http://dx.doi.org/10.1016/B978-0-444-59576-8.00039-4]

[236] Martínez-Medina A, Fernandez I, Lok GB, Pozo MJ, Pieterse CMJ, Van Wees SCM. Shifting from priming of salicylic acid- to jasmonic acid-regulated defences by *Trichoderma* protects tomato against the root knot nematode *Meloidogyne incognita*. New Phytol 2017; 213(3): 1363-77.
[http://dx.doi.org/10.1111/nph.14251] [PMID: 27801946]

[237] Khan RAA, Najeeb S, Hussain S, Xie B, Li Y. Bioactive secondary metabolites from *Trichoderma* spp. against phytopathogenic fungi. Microorganisms 2020; 8(6): 817.
[http://dx.doi.org/10.3390/microorganisms8060817] [PMID: 32486107]

[238] Thambugala KM, Daranagama DA, Phillips AJL, Kannangara SD, Promputtha I. Fungi *vs.* fungi in biocontrol: An overview of fungal antagonists applied against fungal plant pathogens. Front Cell Infect Microbiol 2020; 10: 604923.
[http://dx.doi.org/10.3389/fcimb.2020.604923] [PMID: 33330142]

[239] Palmieri D, Ianiri G, Del Grosso C, *et al.* Advances and perspectives in the use of biocontrol agents against fungal plant diseases. Horticulturae 2022; 8(7): 577.
[http://dx.doi.org/10.3390/horticulturae8070577]

[240] Wepuhkhulu M, Kimenju J, Anyango B, Wachira P. Kyalo. Effect of soil fertility manangement practices and *Bacillus subtilis* on plant parasitic nematodes associated with common bean, *Phaseolus vulgaris*. Trop Subtrop Agroecosystems 2011; 13: 27-34.

[241] Mazzuchelli RCL, Mazzuchelli EHL, Araujo FF. Efficiency of *Bacillus subtilis* for root-knot and lesion nematodes management in sugarcane. Biol Control 2020; 143: 104185.
[http://dx.doi.org/10.1016/j.biocontrol.2020.104185]

[242] Zhao J, Liu D, Wang Y, Zhu X, Chen L, Duan Y. Evaluation of *Bacillus aryabhattai* Sneb517 for control of *Heterodera glycines* in soybean. Biol Control 2020; 142: 104147.
[http://dx.doi.org/10.1016/j.biocontrol.2019.104147]

[243] Antil S, Kumar R, Pathak DV, *et al.* Potential of *Bacillus altitudinis* KMS-6 as a biocontrol agent of *Meloidogyne javanica*. J Pest Sci 2022; 95(3): 1443-52.
[http://dx.doi.org/10.1007/s10340-021-01469-x]

[244] Lee YS, Kim KY. Antagonistic Potential of *Bacillus pumilus* L1 against root-knot nematode, *Meloidogyne arenaria*. J Phytopathol 2016; 164(1): 29-39.
[http://dx.doi.org/10.1111/jph.12421]

[245] Youssef MMA, Abd-El-Khair H, El-Nagdi WMAEH. Management of root knot nematode, *Meloidogyne incognita* infecting sugar beet as affected by certain bacterial and fungal suspensions. Agric Eng Int CIGR J 2017; 2017: 293-301.

[246] Li X, Hu HJ, Li JY, Wang C, Chen SL, Yan SZ. Effects of the endophytic bacteria *Bacillus cereus*

BCM2 on tomato root exudates and *Meloidogyne incognita* infection. Plant Dis 2019; 103(7): 1551-8.
[http://dx.doi.org/10.1094/PDIS-11-18-2016-RE] [PMID: 31059388]

[247]   Ali AAI, El-Ashry RM, Aioub AAA. Animal manure rhizobacteria co-fertilization suppresses phytonematodes and enhances plant production: Evidence from field and greenhouse. J Plant Dis Prot 2022; 129(1): 155-69.
[http://dx.doi.org/10.1007/s41348-021-00529-9]

[248]   Caccia M, Marro N, Dueñas JR, Doucet ME, Lax P. Effect of the entomopathogenic nematode-bacterial symbiont complex on *Meloidogyne hapla* and *Nacobbus aberrans* in short-term greenhouse trials. Crop Prot 2018; 114: 162-6.
[http://dx.doi.org/10.1016/j.cropro.2018.07.016]

[249]   Lewis EE, Grewal PS, Sardanelli S. Interactions between the *Steinernema feltiae–Xenorhabdus bovienii insect* pathogen complex and the root-knot nematode *Meloidogyne incognita*. Biol Control 2001; 21: 55-62.
[http://dx.doi.org/10.1006/bcon.2001.0918]

CHAPTER 2

Evaluation of Damage and Protection in Nematode Infected Plants

Raman Tikoria[1,*], **Roohi Sharma**[2], **Priyanka Saini**[3], **Harsh Gulati**[4] and **Puja Ohri**[2]

[1] *Department of Zoology, School of Basic Sciences, Central University of Punjab, Bathinda 151401, India*

[2] *Department of Zoology, Guru Nanak Dev University, Amritsar 143005, India*

[3] *Department of Zoology, Deshbandhu College, University of Delhi, Delhi 110019, India*

[4] *Department of Zoology, School of Bioengineering and Biosciences, Lovely Professional University, Phagwara, Punjab 144411, India*

Abstract: Nematodes, especially plant-parasitic ones, do a great deal of harm to plants, mostly by attacking the root systems. These tiny roundworms persist in the topmost layer of soil and eat the belowground portion, which prevents the plant from getting the vital nutrients and water it needs. As a result, afflicted plants show signs such as stunted development, which is noticeable even under ideal circumstances, and withering even in the presence of adequate soil moisture. The damage also affects the leaves, which frequently become yellow as a result of nutritional shortages brought on by compromised root function. Reduced yields are frequently the result of damaged root systems that are unable to sustain strong plant development. Additionally, the induction of lesions, galls, and deformities on roots caused by nematode feeding exacerbates the suffering experienced by plants and creates openings for other infections. In severe cases-especially in young or weak specimens - the cumulative effects result in plant death. These consequences highlight the serious threat that nematodes represent to agricultural output, which calls for the application of a number of management techniques to lessen their negative effects and protect crop yields and health. In order to combat nematode infestations, plants have developed a variety of defense systems that include both chemical and physical tactics. To combat nematode infection, plants have developed several defense mechanisms which include both physical and chemical nature. Physical barriers that prevent nematodes from penetrating roots and causing harm include thicker cell walls, lignification, and the creation of suberin layers. In reaction to nematode infestation, plants simultaneously release an abundance of secondary metabolites. These substances, which have nematicidal qualities and directly target nematodes or prevent them from establishing

---

* **Corresponding author Raman Tikoria:** Department of Zoology, School of Basic Sciences, Central University of Punjab, Bathinda 151401, India;
E-mails: raman.tikoria@cup.edu.in, ramankumar404@gmail.com

Shivam Jasrotia & Ajay Kumar (Eds.)

feeding sites, include phytoalexins, phenolics, and terpenoids. Keeping the above mentioned facts in mind, this chapter tends to focus primarily on the damages caused by the plants to their hosts and the nature of defense strategies adopted by them.

Keywords: Allelochemicals, Chlorosis, Defense, Host, Nematodes, Oxidative damage.

INTRODUCTION

With global distribution, root-knot nematodes (RKN) are among the destructive families of plant parasitic nematodes (PPNs) [1]. Among more than 100 species in the genus *Meloidogyne*, "major" species include *M. arenaria, M. incognita, M. hapla,* and *M. javanica* [2]. Due to worries about environmental safety and human health, biological control agents (BCA) are receiving more attention as they appear to represent the future of affordable, environmentally friendly, and sustainable agriculture [3]. These nematodes are often disregarded since most farmers are ignorant of their presence and the symptoms they cause. More than 90% of the anticipated losses are thought to have been caused by the most prevalent species, which impact both field and vegetable crops: *M. arenaria, M. hapla, M. javanica,* and *M. incognita* [4]. The most pervasive and destructive plant parasites are RKN (*Meloidogyne* spp.), which can reduce yield output by 20-50% across large areas of farmed land [5, 6].

In India, these nematodes cause output losses of 16.67, 14.10, 27.21, and 12.3% in tomato, banana, brinjal, and okra, respectively. These worms' attacks worsen the harm by providing a pathway for additional pathogens to infect the host. The deadly, tiny root-knot nematodes cost agriculture $85 billion annually worldwide. In India, it is projected that nematodes cause agricultural losses worth Rs 242.1 billion annually. Worldwide, nematodes still pose a threat to agricultural crops [7]. Up to a 23% decrease in hot pepper production nationwide has been ascribed to RKNs [8]. Merely 10% of all recognized nematode species are harmful to plants, meaning they feed on plants, which hinders crop growth and production efficiency. In tropical and subtropical regions, PPNs have been predicted to bring about a 14.6% reduction in agricultural output, whereas in industrialized regions, the loss is 8.8%. According to study [9], PPNs cause around 10% of the global crop output losses that occur each year or $173 billion in US dollars. The availability of food is significantly impacted by PPNs because of the sharp growth in food demand brought on by the expected 35% growth in the number of humans by 2050. According to some estimates, dietary trends and constant increases in the population would lead to a 75% increase in food consumption between 2010 and 2050 [10]. Studies show that tomato production is decreased by RKN by 26.5% to 73.3%, costing the global tomato industry around $125 billion annually in losses

[11]. Generally speaking, foliar indications of nematode infection include nitrogen deficiencies and tomato production losses due to root-knot nematodes.

According to a study [12], RKN rank the top among the 10 most significant genera of PPNs as well as first among the five primary plant illnesses. They are widely distributed geographically, have a wider host range, and have a potent damaging ability. They have been connected to a number of plant illnesses that aggravate wilt diseases and create disease complexes [13].

NEMATODE PATHOGENESIS IN PLANTS

The phylum Nematoda contains nematodes, which are tiny creatures that may be found in a range of environments such as freshwater, aquatic, soil, and marine. They can be free-living organisms or parasites on bacteria, fungi, plants, or mammals. Due to their widespread occurrence in nature, they pose a threat to humans, animals, and plants. They also cause losses in agricultural productivity on a worldwide scale [14, 15]. Their eating habits and surroundings are the main factors used to categorize them. In contrast to ectoparasites, which feed through stylets inside root cells, endoparasites invade roots. Sedentary parasites, RKNs grow knot-like forms on root surfaces and develop specialized organs to get nutrients from their hosts [16]. In addition, they cause damage to the plants by decreasing water absorption and mineral availability due to disturbances in root architecture. Additionally, they increase the host plants' susceptibility to disease attacks [17].

At the elongation zone, the juveniles in their second stage (J2) penetrate the plant roots and break down the cell wall using enzymes that degrade cell walls, like glucanases, pectinases, and chitinases [18]. They go into the vascular zone from the apoplast region, where they settle in and create large numbers of feeding cells. These enormous cells are assumed to be the precursors of hyperplasic tissues and gall formation. In response to nematode effectors, plants disrupt the molecular pathways that link them [19]. Repetitive mitosis, cytokinesis, and duplication are partially completed by the large cells, leading to incomplete division or mitosis and the amplification of DNA necessary for giant cell proliferation [20]. Galls arise once asynchronous development is complete, and this is intimately linked to the infectious stage of RKN [21]. Additionally, giant cells form in the vicinity of cell wall zones, producing asymmetric membrane thickenings that facilitate solute transport and cell invasions (Fig. **1**).

Stunted Growth

Nematodes damage the root system, hindering the capacity of plants to take up water and nutrients from the soil, which causes stunted growth and poor overall development [22]. Through a variety of processes that impair the proper operation of the root system, impede nutrient uptake, and impede plant development, nematodes can cause stunted growth in plants. Nematodes mostly cause stunted development by feeding on the roots of plants [23]. With their stylets, specialized mouthparts are used for feeding plant-parasitic nematodes that puncture the root surface and withdraw or consume the contents of the cells through their oral apertures. This feeding activity wears down the root tissues, resulting in necrosis and the development of lesions or galls that obstruct the host's capacity to absorb nutrients, water, and signaling chemicals. Moisture stress, nutritional shortages, and stunted development ensue from the afflicted roots' decreased water and vital minerals uptake from the soil [24].

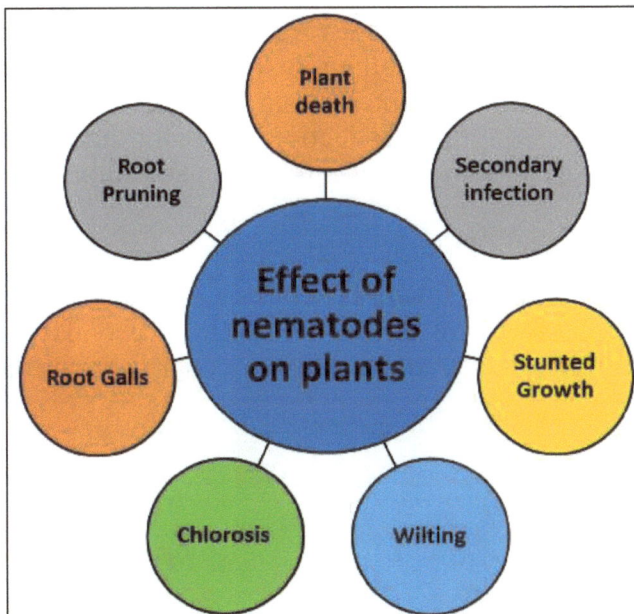

Fig. (1). Effect of nematodes on plants.

Wilting

Plant-parasitic nematodes cause water stress and wilting signs in the plant because the impacted roots are unable to sufficiently give water to the sections of the plant above ground [25]. Furthermore, feeding nematodes can cause internal modulations in the plant, like the release of hormones linked to stress, which worsen wilting and water stress. Nematodes may potentially cause harm to the root's vascular tissues, which would impede the host's capacity to transmit nutrients and water [26].

Yellowing (Chlorosis)

Plants can get yellowing, or chlorosis, from nematodes through a variety of processes that interfere with the plant's capacity to intake and use vital nutrients [27]. Nematodes harm the root system, which is one of the main ways they produce chlorosis. Plant-parasitic nematodes physically harm plants by consuming their roots and changing the physiology of the belowground part [28]. Damage like this can make it more difficult for the roots to take in water and nutrients from the soil, which can result in deficits in important nutrients like iron, potassium, and nitrogen [29]. Plants that get insufficient amounts of these nutrients may develop chlorosis, a condition in which the green pigment chlorophyll—which is necessary for photosynthesis-is not produced in sufficient amounts [30]. Furthermore, nematodes can obstruct the intake of nutrients by causing galls or lesions to grow on the roots, which obstruct the plant's ability to transfer nutrients. Chlorosis symptoms can also be exacerbated by nematode feeding, which can cause the production of plant hormones and signaling molecules that change metabolism and resource allocation [31].

Root Lesions and Deformities

Some nematode species cause visible lesions or galls on the roots. These lesions can provide entry points for pathogens, leading to secondary infections. Galls, which are abnormal growths on roots, can disrupt normal root function [32].

Root Pruning, Lesions and Deformities

Nematodes feed on root tissue, resulting in the pruning of roots. Severely pruned roots cannot adequately support the plant, leading to further stress and susceptibility to other stressors such as drought or disease [32, 33].

Secondary Infections

Weakened plants are highly vulnerable to secondary infections by fungi or bacteria. Nematodes injure the root tissue and produce sores or lesions when they feed on plant roots [34]. These cuts provide openings for bacteria and fungus, two types of soil-borne diseases, to enter the plant tissues. Furthermore, nematode feeding may cause physiological alterations in the plant, such as the production of chemicals linked to stress or changes in hormone levels, which may weaken the plant's defenses against infection and increase its susceptibility to further infections [35]. For instance, it is known that some nematodes of *Meloidogyne* spp. cause distinctive galls to grow on plant roots. These galls provide a safe sanctuary for different diseases to flourish and infect the plant, including fungi that cause root rot (such as *Fusarium* and *Pythium* species) [36]. Similar to this, opportunistic pathogens can colonize lesions caused by lesion worms (*Pratylenchus* spp.) feeding on root tissues, resulting in necrosis and subsequent infections [32]. Moreover, the plant's general health and vitality may be compromised by nematode-induced stress, increasing its susceptibility to illness. For instance, signs of nutrient shortage, stunted growth, or wilting may be seen in nematode-infested plants, which may reduce their resistance to disease invasions. Since nematode infestations already cause stress to plants, they may increase their receptivity to invasion by getting more infections found in the soil or surrounding environment [25].

Death of Plants

In severe cases of nematode infestation, particularly in young or stressed plants, death can occur due to the extensive damage to the foundational framework and the inability of the host to absorb water and nutrients effectively [25]. Table **1** summarizes various studies that report the effect of nematodes on plant growth and its related phytoconstituents.

Table 1. Effect of nematode stress on plant growth and related phytoconstituents.

Sr. No.	Plant	Nematode Species	Result	References
1.	*Trachyspermum ammi*	*M. incognita*	The activity of CAT and POD increased with nematode infection after 120 days.	[56]
2.	Tomato	*M. incognita*	An average of 997 galls and 842 mean egg masses were observed in plants inoculated with 2000 J2s.	[57]
3.	Tomato	*M. incognita*	Increased level of antioxidative enzymes, increased level of MDA, H_2O_2, decreased level of photosynthetic pigments, and increased level of proline along with other phenolic compounds after nematode infection.	[58]
4.	Tomato	*M. incognita*	Root fresh weight, activities of SOD, POD, CAT, APOX, DHAR, GST, GR, and PPO was enhanced by 32.1, 64.7%, 45.2%, 58.0%, 55.2%, 38.8%, 39.6%, 59.6%, and 86.4%, respectively. Following nematode infection, there was a 46% reduction in protein content.	[59]
5.	Tomato	*M. incognita*	Chlorophyll a, chlorophyll b, and carotenoid levels were reduced by 10, 18 and 20%, respectively, after nematode infection.	[60]
6.	Tomato	*M. incognita*	Approximately 2250 juveniles and eggs were found per gram of root, with a root gall index of about 5.5.	[61]

(Table 1) cont.....

Sr. No.	Plant	Nematode Species	Result	References
7.	Tomato	*M. incognita*	The nematode-infected plants had more peroxide activity than the non-infected plants.	[62]
8.	Tomato	*M. incognita*	Root length, shoot length, fresh weight, dry weight, total chlorophyll, photosynthetic rate, stomatal conductance, intercellular CO_2, and transpiration rate reduced by 43.2, 21.5, 35.1, 47.6, 19.7, 33.9, 31.6, 11.3 and 40.3%. Further, SOD, POD, CAT, APOX, GPOX, DHAR, GST, PPO activities MDA, and H_2O_2 content enhanced by 34,100, 31, 13, 42, 9, 26, 19, 168, and 195%, respectively.	[63]
9.	Tomato	*M. incognita*	Chlorophyll decreased (10–20%) and carotenoid levels (16–20%) and an increase in MDA was observed following nematode infection.	[64]
10.	Cucumber	*M. incognita*	Two-week-old plants with an inoculum density of 8000 J2s per plant have lower plant height, lower fruit output, and more galls.	[65]
11.	Cucumber	*M. incognita*	There were significant variations between all cucumber cultivars regarding gall development, egg masses, fecundity, and reproductive factors.	[66]
12.	Tomato	*M. incognita*	Root length, dry weight total chlorophyll, Chlorophyll a and b were reduced by 21.09, 42.74, 47.4, 48.16, and 43.96%. Ascorbic acid, glutathione, tocopherol, and carbohydrate contents were enhanced by 18.75, 35.56, 52.94 and 65.22%.	[67]
13.	*Arabidopsis thaliana*	*Heterodera schachtii*	H_2O_2 level, the activity of antioxidative enzymes, proline, and phenolic compounds increase with nematode infection.	[68]
14.	Tomato	*M. incognita*	There was a 10 and 51% reduction in tomato plant height and shoot fresh weight, respectively.	[69]
15.	Tomato	*M. incognita*	Elevated levels of MDA and H_2O_2 by 2.47 and 1.63 times, and 49 and 50% lower levels of chlorophyll a and b in tomatoes infected with nematodes.	[70]
16.	*Abelmoschus esculentus*	*M. incognita*	Nematode infection is directly proportional to the inoculum density with a reduction in plant height and biomass.	[71]
17.	*Abelmoschus esculentus*	*M. incognita*	On highly sensitive individuals, a 17.1-fold reproductive factor, average egg masses (177.8), and eggs per egg mass (298.4) were reported.	[72]
18.	Tomato	*M. arenaria*	Substantial reductions in plant height, fresh weight of the roots, and fresh weight of the shoots.	[73]
19.	Cotton	*M. incognita*	Decrease in chlorophyll concentration.	[74]

(Table 1) cont.....

Sr. No.	Plant	Nematode Species	Result	References
20.	Black gram	*M. incognita*	The shoot's fresh and dry weights were drastically decreased.	[75]
21.	*Sphenostylis stenocarpa*	*M. incognita*	Juveniles have maximum gall intensity and reduced plant height as compared to those who were uninfected.	[76]
22.	Tomato	*M. ethiopica*	The photosynthetic rate is reduced by 60–70%.	[77]
23.	*Morinda citrifolia*	*M. incognita*	About 42-47% yield loss was observed in two successive seasons.	[78]
24.	*Lens culinaris*	*M. incognita*	*Lens culinaris* roots and shoots with an inoculum density of 500–4000 J2s per treatment showed 69 and 50% decreases in root and shoot length and 81 and 73% reduction in root and shoot dry weight, respectively.	[79]
25.	Black gram	*M. incognita*	In comparison to non-inoculated plants, plant height, or root and shoot length (14%), fresh matter (11%), and dry matter (11%) all decreased following nematode infection.	[80]
26.	*Vigna radiata*	*M. javanica*	Plants infected with nematodes have lower levels of chlorophyll and carotenoids. For up to 30 days after inoculation, phenols increased in the leaves compared to the controls. After 30 days of nematode exposure, protein levels dropped considerably.	[81]
27.	Black gram	*M. incognita*	Reduction in the length of the root and shoot.	[82]
28.	*Phaseolus vulgaris*	*M. incognita* and *M. javanica*	Untreated plants have more gall index as compared to those that are supplemented with green plant manure.	[83]
29.	*Trigonella foenum-graecum*	*M. incognita*	The life cycle completes within 34 days in susceptible varieties whereas it takes 42-52 days in resistant varieties.	[47]

PLANT RESPONSE TO NEMATODE INFECTION

The nematode's style, is a spear-like device attached to its mouth that inserts itself into plant tissue to cause parasitism [37]. It is a spear-like device attached to its mouth, and inserts itself into plant tissue to cause parasitism [38]. Dirt hides the infection, even if it happens in the root system of the host. Therefore, the reaction symptoms observed in the shoot provide direction for determining the nematode prevalence in the area. *Psidium guajava* L. was shown to be experiencing visual signs such as yellow or dark red patches, death, the early collapse of leaves and withering, and strength loss as a result of *Meloidogyne enterolobii* infection [39].

Due to the significant damage that these organisms may inflict in order to protect food supplies and prevent financial losses, nematode-controlling drugs, and test items were first produced in the eighteenth century. Root-knot nematode infections can cause oval patches, galling of the roots, stunting, chlorosis, early wilting, and a general decline in plant growth in addition to stunted or delayed development. Plants that are infected have a yellowish appearance, and the infection causes root galling and wilting [40].

The capacity of a plant to take up both nutrients and water is compromised, which lowers tomato quality and output. Environmental factors, nematode species, soil type, degree of soil infection, and cultivar all influence damage [41]. Plants under nematode stress have altered antioxidant activity, mostly due to the production of free radicals [42]. Plant defense mechanisms against nematode infection have been highlighted in Fig. (**2**).

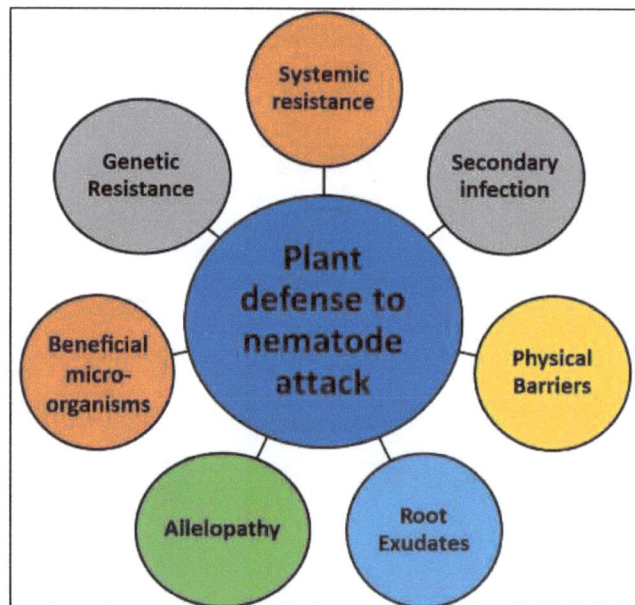

Fig. (2). Plant defense mechanism against nematode infection.

Physical Barriers

Plants can deploy mechanical obstacles such as thickened cell walls, lignin formation, and the formation of suberin layers to restrict nematode penetration into roots. These barriers can hinder nematode movement and feeding, reducing the extent of damage [43]. Nematodes trying to enter plant roots face a strong barrier made up of thickened cell walls, which hinders their movement and restricts their access to nutrient-rich root tissues. Furthermore, lignin offers architectural assistance and resilience to mechanical stress and may be used by plants to strengthen their cell walls [44]. Lignification fortifies root tissues even more, increasing their resistance to harm caused by nematodes. The creation of suberin layers, which function as impermeable barriers that hinder nematode mobility and their capacity to establish feeding sites within roots, is another physical barrier strategy used by plants. Plants may effectively fight nematode invasion and lessen the chance of substantial root system damage by strengthening their root tissues with these physical barriers [43].

Induced Resistance

Upon nematode attack, plants can trigger Systemic Acquired Resistance (SAR) or Induced Systemic Resistance (ISR) mechanisms. These responses involve the emergence of defense-related and signaling compounds that improve the plant's resistance to nematode infestation. In the end, these induced defenses lessen the detrimental impacts of nematode infestation on host health by directly targeting nematodes, interfering with their feeding habits, or preventing their reproduction. For instance, upon nematode sensing, plants may release poisonous chemicals or secondary metabolites that are harmful to nematodes. In order to prevent nematodes from penetrating plant tissues, physical barriers like lignin deposition or cell wall strengthening can also be activated as a means of induced resistance [35]. Furthermore, the synthesis of salicylic acid and jasmonic acid, which coordinate the plant's systemic defensive responses and increase its resistance to nematode assaults, can be stimulated by induced resistance. Induced resistance benefits the plant by preparing its immune system to respond forcefully and quickly to nematode invasion [24].

Root Exudates

Through their root systems, plants emit a wide range of substances into the soil, such as organic acids, carbohydrates, and proteins [45]. Some of these compounds can attract beneficial microbes that antagonize nematodes or induce systemic resistance in the plant. Root exudates are essential for reducing or eliminating nematode damage to plants because they affect nematode behavior, encourage the growth of advantageous microbes, and stimulate plant defenses [25]. A few of these substances directly impact nematodes, either by repelling them, limiting their movement, or obstructing their capacity to find host plants. For instance, certain root exudates can interfere with nematode chemotaxis, which is how worms find their way to plant roots, lowering the attractiveness and infection rates of nematodes [46]. Moreover, the growth of advantageous microorganisms in the rhizosphere, such as rhizobacteria that stimulate plant growth and mycorrhizal and nematode-trapping fungi, can be encouraged by root exudates. These microorganisms possess the capacity to repel nematodes and elicit fundamental resilience in plants. These helpful microbes have the ability to create substances that are harmful to nematodes, outcompete plant-parasitic nematodes for resources, or activate plant defense mechanisms in response to nematode infection [47]. Root exudates reduce the chance of nematode harm to plants by altering the makeup and activity of the soil microbiome, which makes the environment less conducive to nematode survival and reproduction. All things considered, root exudates are an essential part of plant-nematode interactions and provide opportunities for long-term nematode control in agriculture [48].

Allelopathy

Some plants release allelochemicals into the soil that inhibit nematode activity or disrupt their life cycle [25]. These compounds can be synthesized by the plant itself or produced by neighboring plants, contributing to intercropping or crop rotation strategies for nematode management [49]. For instance, it has been demonstrated that some plants generate allelochemicals such as terpenoids, phenolics, and flavonoids that have nematode-killing action against different species. This infestation is able to possess a less detrimental impact on the well-being of plants when these allelochemicals interfere with nematode physiology, prevent them from reproducing, or even cause them to die [50]. Furthermore, the content and structure of the soil microbiome can be changed by allelopathic plants, promoting the growth of advantageous bacteria that irritate worms or stimulate plant defenses against nematode infection [51].

Beneficial Microorganisms

Plants can form symbiotic relationships with useful microorganisms like rhizobacteria and mycorrhizal fungi, which can indirectly suppress nematode populations by enhancing plant health and activating defense mechanisms [52]. Through a number of processes, beneficial microbes act as plants' natural friends to defend them from nematode infection. As a whole, they support sustainable pest control practices in agriculture. Among these microbes, certain predatory nematodes, like those in the genus *Pratylenchus*, actively hunt down plant-parasitic nematodes to lower their numbers in the soil and lessen the chance that plants would get infected [32]. Nematophagous fungi, such as *Arthrobotrys* and *Paecilomyces*, also create structures that catch nematodes, thereby limiting their population in the soil and shielding plant roots. Examples of these structures include constricting rings and sticky traps. Furthermore, bacteria that trap nematodes, including *Bacillus* and *Pseudomonas*, have the ability to create substances that are poisonous to nematodes or to compete with them for resources in the rhizosphere, which lowers nematode populations and improves plant health

[52]. Through their symbiotic associations with plant roots, mycorrhizal fungi enhance nutrient intake and development while simultaneously promoting systemic resistance against worm infection. Last but not the least, rhizosphere-dwelling plant growth-promoting rhizobacteria, like *Bacillus* and *Pseudomonas*, enhance plant defenses by either producing antimicrobials or inducing systemic resistance to nematodes. When taken as a whole, these advantageous microbes offer viable pathways for the biological management of nematode pests, providing sustainable and green ways to safeguard crops against nematode harm [53].

Genetic Resistance

Plant breeding programs have developed cultivars with genetic resistance to specific nematode species or strains [54]. These resistant varieties often possess genes that confer tolerance to nematode infection, reducing the severity of damage or preventing infestation altogether. Certain plants have developed defense mechanisms, such as the creation of poisonous chemicals or the thickening of cell walls to prevent nematode penetration, in response to nematode invaders. By preventing nematode growth or reproduction within plant roots, some resistance genes lower nematode populations and lessen the amount of harm they cause. Additionally, genetic resistance may result in altered root morphology, such as thicker or more branched root hairs, which may make it more difficult for nematodes to find feeding sites. Genetically, plants may be able to recognize nematodes in the soil and set up defenses against or avoidance strategies before an infestation takes place [55]. Soybean cultivars created to fight the soybean cyst nematode (SCN), among the most economically destructive pests impacting soybean crops globally, are one example of plants that have gained genetic resistance to nematodes. Using conventional breeding techniques, wild-derived resistance genes of soybean relatives were included in the development of SCN-resistant soybean cultivars. The Rhg1 locus, which has many genes that give resistance to various SCN populations, is the main source of resistance in soybeans. These genes inhibit nematode populations in the soil, decrease cyst development, and restrict nematode reproduction within the roots. For example, breeding efforts have made extensive use of the Rhg1b allele found in the PI 88788 soybean genotype to create SCN-resistant soybean cultivars. The Peking genotype is another source of resistance, which carries the Rhg4 gene [54].

CONCLUSION

Intensive agriculture contributes to large agricultural yields but can also create major environmental consequences. Specifically, the overuse of fertilizers and pesticides has seriously harmed human health and the environment. There are around 3400 different kinds of nematodes that may parasitize plants, reducing crop productivity and causing financial losses. The need for food rises in tandem with the growth of the human population. They harm the flora in several ways, including stunted growth, wilting, and death of plants, and cause huge harm to the agricultural sector. In order to fend off nematode infections, plants have developed a variety of defense strategies, including chemical and physical barriers, induced resistance, root exudates, allelopathy, symbiotic relationships with helpful microorganisms, and genetic resistance. Together, these defensive mechanisms lessen the adverse consequences of nematode infestations on the production and wellness of plants. Although every mechanism has advantages and disadvantages of its own, combining many methods into comprehensive nematode management plans offers hope for long-term control. Researchers and farmers may create efficient plans to shield crops from nematode damage while advancing agricultural sustainability and food security by comprehending the intricate relationships between plants and nematodes and utilizing a variety of defensive mechanisms.

REFERENCES

[1]   Jones JT, Haegeman A, Danchin EG, *et al.* Top 10 plant-parasitic nematodes in molecular plant pathology. Mol Plant Pathol 2013; 14(9): 946-61.
       [PMID: 23809086]

[2]   Elling AA. Major emerging problems with minor *Meloidogyne* species. Phytopathology 2013; 103(11): 1092-102.
       [http://dx.doi.org/10.1094/PHYTO-01-13-0019-RVW] [PMID: 23777404]

[3]   Rao MS, Kamalnath M, Umamaheswari R, *et al. Bacillus subtilis* IIHR BS-2 enriched vermicompost controls root knot nematode and soft rot disease complex in carrot. Sci Hortic (Amsterdam) 2017; 218: 56-62.

[4]   Marquez J, Hajihassani A. Identification, diversity, and distribution of *Meloidogyne* spp. in vegetable fields of South Georgia, USA. Phytopathology 2023; 113(6): 1093-102.
       [PMID: 36449528]

[5]   Lamovšek J, Gregor UR, Trdan S. Biological control of root-knot nematodes (*Meloidogyne* spp.): microbes against the pests. Acta Agric Slov 2013; 101(2): 263-75.

[6]   Khan A, Ahmad G, Haris M, Khan AA. Bio-organics management: Novel strategies to manage root-knot nematode, *Meloidogyne incognita* pest of vegetable crops. Gesunde Pflanzen 2023; 75(1): 193-209.

[7]   Thakur RK, Shirkot P. Potential of biogold nanoparticles to control plant pathogenic nematodes. J Bioanal Biomed 2017; 9(4): 220-2.

[8]   Khan MR, Jain RK, Ghule TM, Pal S. Root knot nematodes in India. A comprehensive monograph. All india co-ordinated research project on plant parasitic nematodes with integrated approach for their control. Indian Agricultural Research Institute, New Delhi, PP78. 2014; 29.

[9]   Dutta TK, Khan MR, Phani V. Plant-parasitic nematode management *via* biofumigation using brassica and non-brassica plants: Current status and future prospects. Curr Plant Biol 2019; 17: 17-32.

[10]   Tilman D, Balzer C, Hill J, Befort BL. Global food demand and the sustainable intensification of agriculture. Proc Natl Acad Sci USA 2011; 108(50): 20260-4.
       [PMID: 22106295]

[11]   Rawal S. A review on root-knot nematode infestation and its management practices through different approaches in tomato. Trop Agroecosystem 2020; 1: 92-6.

[12]   Mukhtar T, Arooj M, Ashfaq M, Gulzar A. Resistance evaluation and host status of selected green gram germplasm against *Meloidogyne incognita.* Crop Prot 2017; 92: 198-202.

[13]   Aslam MN, Mukhtar T, Ashfaq M, Hussain MA. Evaluation of chili germplasm for resistance to bacterial wilt caused by *Ralstonia solanacearum.* Aus Plant Pathol 2017; 46: 289-92.

[14]   Tikoria R, Kaur A, Ohri P. Potential of vermicompost extract in enhancing the biomass and bioactive components along with mitigation of *Meloidogyne incognita*-induced stress in tomato. Environ Sci Pollut Res Int 2022; 29(37): 56023-36.
[PMID: 35332451]

[15]   Tikoria R, Kaur A, Ohri P. Modulation of various phytoconstituents in tomato seedling growth and *Meloidogyne incognita*–induced stress alleviation by vermicompost application. Front Environ Sci 2022; 10: 891195.

[16]   Favery B, Dubreuil G, Chen MS, Giron D, Abad P. Gall-inducing parasites: Convergent and conserved strategies of plant manipulation by insects and nematodes. Annu Rev Phytopathol 2020; 58: 1-22.
[PMID: 32853101]

[17]   Trudgill DL. Resistance to and tolerance of plant parasitic nematodes in plants. Annu Rev Phytopathol 1991; 29(1): 167-92.

[18]   Tikoria R, Kaur A, Ohri P. Amelioration of oxidative stress and growth enhancement by application of vermicompost *via* modulating phyto-constituents in tomato plants during nematode stress. J Soil Sci Plant Nutr 2023; 23(3): 3944-60.

[19]   Truong HD, Wang CH, Kien TT. Effect of vermicompost in media on growth, yield and fruit quality of cherry tomato (*Lycopersicon esculentun* Mill.) under net house conditions. Compost Sci Util 2018; 26(1): 52-8.

[20]   de Almeida Engler J, Vieira P, Rodiuc N, Grossi de Sa MF, Engler G. The plant cell cycle machinery: Usurped and modulated by plant-parasitic nematodes. In: Advances in botanical research 2015; 73: 91-118.
[http://dx.doi.org/10.1016/bs.abr.2014.12.003]

[21]   Escobar C, Barcala M, Cabrera J, Fenoll C. Overview of root-knot nematodes and giant cells. In: Advances in botanical research 2015; 73: 1-32.
[http://dx.doi.org/10.1016/bs.abr.2015.01.001]

[22]   Tikoria R, Kumar D, Ali M, Ohri P. Boosting of free radical scavenging capacity and physiological markers of tomato plants by vermicompost application during nematode stress. J Soil Sci Plant Nutr 2024; 24(1): 1507-18.
[http://dx.doi.org/10.1007/s42729-024-01656-6]

[23]   Kumar Y, Yadav BC. Plant-parasitic nematodes: Nature's most successful plant parasite. Int J Res Rev 2020; 7(3): 379-86.

[24]   Ali M, Kumar D, Tikoria R, *et al.* Exploring the potential role of hydrogen sulfide and jasmonic acid in plants during heavy metal stress. Nitric Oxide 2023; 140-141: 16-29.
[PMID: 37696445]

[25]   Sikder MM, Vestergård M. Impacts of root metabolites on soil nematodes. Front Plant Sci 2020; 10: 1792.
[http://dx.doi.org/10.3389/fpls.2019.01792] [PMID: 32082349]

[26]   Gillet FX, Bournaud C, Antonino de Souza Júnior JD, Grossi-de-Sa MF. Plant-parasitic nematodes: towards understanding molecular players in stress responses. Ann Bot (Lond) 2017; 119(5): 775-89.
[PMID: 28087659]

[27]   El-Sagheer AM. Plant responses to phytonematodes infestations In: Ansari RA, Mahmood I Eds, Plant health under biotic stress Microbial Interactions 2019; 2: 161-75.
[http://dx.doi.org/10.1007/978-981-13-6040-4_8]

[28]   Tikoria R, Sharma N, Kour S, Kumar D, Ohri P. Vermicomposting: an effective alternative in integrated pest management. In: Vig AP, J. Singh J, Suthar S, Eds, Earthworm Engineering and Applications. New York: Nova Science Publishers 2022c; pp. 103-118.

[29]   de Melo Santana-Gomes S, Dias-Arieira CR, Roldi M, Santo Dadazio T, Marini PM, de Oliveira Barizatilde DA. Mineral nutrition in the control of nematodes. Afr J Agric Res 2013; 8(21): 2413-20.

[30]   Tikoria R, Ohri P. Application of neem waste vermicompost in compensating nematode induced stress and upregulating physiological markers of tomato plants under glass house conditions after 10 days of exposure. Environ Sci Pollut Res Int 2023; 1-1.
[http://dx.doi.org/10.1007/s11356-023-30324-y] [PMID: 37864696]

[31]   Kyndt T, Fernandez D, Gheysen G. Plant-parasitic nematode infections in rice: molecular and cellular insights. Annu Rev Phytopathol 2014; 52(1): 135-53.
[http://dx.doi.org/10.1146/annurev-phyto-102313-050111] [PMID: 24906129]

[32]   Jones MGK, Fosu-Nyarko J. Molecular biology of root lesion nematodes (*Pratylenchus* spp.) and their interaction with host plants. Ann Appl Biol 2014; 164(2): 163-81.
[http://dx.doi.org/10.1111/aab.12105]

[33]   Arieira GO. Diversidade de nematoides em sistemas de culturas e manejo do solo. Universidade Estadual de Londrina 2012.

[34]   Gorshkov V, Tsers I. Plant susceptible responses: the underestimated side of plant–pathogen interactions. Biol Rev Camb Philos Soc 2022; 97(1): 45-66.
[http://dx.doi.org/10.1111/brv.12789] [PMID: 34435443]

[35]   Tikoria R, Sharma N, Kour S, *et al.* Cotton oilseed cake: Chemical composition and nematicidal potential. In: Ahmad F, Pandey R, Eds. Oilseed cake for nematode management. USA: CRC Press 2023; pp. 59-69.
[http://dx.doi.org/10.1201/9781003319252-4]

[36]   Gahukar RT. Management of pests and diseases of important tropical/subtropical medicinal and aromatic plants: A review. J Appl Res Med Aromat Plants 2018; 9: 1-18.
[http://dx.doi.org/10.1016/j.jarmap.2018.03.002]

[37]   Lopez-Nicora H, Peng D, Saikai K, Rashidifard M. Nematode problems in maize and their sustainable management. In: Khan MR, Quintanilla M, Eds, Nematode diseases of crops and their sustainable management. London: Academic Press 2023; 1: 167-81.
[http://dx.doi.org/10.1016/B978-0-323-91226-6.00018-3]

[38]   Lambert K, Bekal S. Introduction to plant-parasitic nematodes. The plant health instructor. 2002.
[http://dx.doi.org/10.1094/PHI-I-2002-1218-01]

[39]   Almeida EJ, Santos JM, Martins ABG. Influência do parasitismo pelo nematoide de galhas nos níveis de nutrientes em folhas e na fenologia de goiabeira 'Paluma'. Bragantia 2011; 70(4): 876-81.
[http://dx.doi.org/10.1590/S0006-87052011000400021]

[40]   Ansari T, Asif M, Siddiqui MA. Potential of botanicals for root-knot nematode management on tomato: root-knot nematode *Meloidogyne incognita* management through organic amendment. New York: LAP LAMBERT Academic Publishing 2016.

[41]   Bakr RA, Mahdy ME, Mousa ES, Salem MA. Efficacy of *Nerium oleander* Leaves extract on controlling *Meloidogyne incognita in-Vitro* and *in-Vivo*. Egypt. J Agric Res (Lahore) 2015; 10(1): 1-3.

[42]   Chawla NC, Kavita Choudhary KC, Sukhjeet Kaur SK, Salesh Jindal SJ. Changes in antioxidative enzymes in resistant and susceptible genotypes of tomato infected with root-knot nematode (*Meloidogyne incognita*). Indian J Nematol 2013; 43(1): 1-12.

[43]   Lamalakshmi Devi E, Kumar S, Basanta Singh T, *et al.* Adaptation strategies and defence mechanisms of plants during environmental stress. In: Ghorbanpour M, Varma A, Eds. Medicinal plants and environmental challenges. Cham: Springer 2017; pp. 359-413.
[http://dx.doi.org/10.1007/978-3-319-68717-9_20]

[44]   Yadav S, Chattopadhyay D. Lignin: the building block of defense responses to stress in plants. J Plant Growth Regul 2023; 42(10): 6652-66.
[http://dx.doi.org/10.1007/s00344-023-10926-z]

[45]   Khan N, Ali S, Zandi P, *et al.* Role of sugars, amino acids and organic acids in improving plant abiotic stress tolerance. Pak J Bot 2020; 52(2): 355-63.
[http://dx.doi.org/10.30848/PJB2020-2(24)]

[46]   Abd-Elgawad MMM. Understanding molecular plant–nematode interactions to develop alternative approaches for nematode control. Plants 2022; 11(16): 2141.
[http://dx.doi.org/10.3390/plants11162141] [PMID: 36015444]

[47]   Sharma A, Trivedi PC. Studies on the life cycle of *Meloidogyne incognita* in two cultivars of *Trigonella foenum-graecum* [India]. Nematol Mediterr, (Italy). 1992; 20(2).

[48]   Topalović O, Vestergård M. Can microorganisms assist the survival and parasitism of plant-parasitic nematodes? Trends Parasitol 2021; 37(11): 947-58.
[http://dx.doi.org/10.1016/j.pt.2021.05.007] [PMID: 34162521]

[49]   Abd-Elgawad M. Optimizing safe approaches to manage plant-parasitic nematodes. Plants 2021; 10(9): 1911.
[http://dx.doi.org/10.3390/plants10091911] [PMID: 34579442]

[50]   Alam EA, Nuby AS. Phytochemical and nematicidal screening on some extracts of different plant parts of Egyptian *Moringa oleifera* L. Pak J Phytopathol 2022; 34(2): 293-306.
[http://dx.doi.org/10.33866/phytopathol.034.02.0818]

[51]   Rathore D, Naha N, Singh S. Phenolic compounds and nanotechnology: application during biotic stress management in agricultural sector and occupational health impacts. In: Lone R, Khan S, Al-Sadi AM Eds, Plant Phenolics in Biotic Stress Management. Singapore: Springer Nature Singapore 2024; pp. 503-49.
[http://dx.doi.org/10.1007/978-981-99-3334-1_21]

[52]   Poveda J, Abril-Urias P, Escobar C. Biological control of plant-parasitic nematodes by filamentous fungi inducers of resistance: *Trichoderma*, mycorrhizal and endophytic fungi. Front Microbiol 2020; 11: 992.
[http://dx.doi.org/10.3389/fmicb.2020.00992] [PMID: 32523567]

[53]   Singh G, Pujari M. *Bacillus subtilis* as a plant-growth-promoting rhizobacteria: a review. Plant Arch (09725210). 2022; 22(2).

[54]   Kim KS, Vuong TD, Qiu D, *et al.* Advancements in breeding, genetics, and genomics for resistance to three nematode species in soybean. Theor Appl Genet 2016; 129(12): 2295-311.
[http://dx.doi.org/10.1007/s00122-016-2816-x] [PMID: 27796432]

[55]   Seid A, Imren M, Ali MA, Toumi F, Pauliitz T, Dababat AA. Genetic resistance of wheat towards plant-parasitic nematodes: Current status and future prospects. Biotech Studies 2021; 30(1): 43-62.
[http://dx.doi.org/10.38042/biotechstudies.944678]

[56]   Danish M, Robab MI, Marraiki N, *et al.* Root-knot nematode *Meloidogyne incognita* induced changes in morpho-anatomy and antioxidant enzymes activities in *Trachyspermum ammi* (L.) plant: A microscopic observation. Physiol Mol Plant Pathol 2021; 116: 101725.
[http://dx.doi.org/10.1016/j.pmpp.2021.101725]

[57]   Seman A, Awol S, Mashilla D. Integrated management of *Meloidogyne incognita* in tomato (*Solanum lycopersicum*) through botanical and intercropping. Afr J Agric Res 2020; 15(4): 492-501.
[http://dx.doi.org/10.5897/AJAR2019.14040]

[58]   Khajuria A, Ohri P. Polyamines induced nematode stress tolerance in *Solanum lycopersicum* through altered physico-chemical attributes. Physiol Mol Plant Pathol 2020; 112: 101544.
[http://dx.doi.org/10.1016/j.pmpp.2020.101544]

[59]   Bali S, Kaur P, Jamwal VL, *et al.* Seed priming with jasmonic acid counteracts root knot nematode infection in tomato by modulating the activity and expression of antioxidative enzymes. Biomolecules 2020; 10(1): 98.
[http://dx.doi.org/10.3390/biom10010098] [PMID: 31936090]

[60]   Udalova ZV, Folmanis GE, Fedotov MA, *et al.* Effects of silicon nanoparticles on photosynthetic pigments and biogenic elements in tomato plants infected with root-knot nematode *Meloidogyne incognita.* Dokl Biochem Biophys 2020; 495(1): 329-33.
[http://dx.doi.org/10.1134/S1607672920060150] [PMID: 33368045]

[61]   D'Addabbo T, Laquale S, Perniola M, Candido V. Biostimulants for plant growth promotion and sustainable management of phytoparasitic nematodes in vegetable crops. Agronomy (Basel) 2019; 9(10): 616.
[http://dx.doi.org/10.3390/agronomy9100616]

[62]   Afifah EN, Murti RH, Nuringtyas TR. Metabolomics approach for the analysis of resistance of four tomato genotypes (*Solanum lycopersicum* L.) to root-knot nematodes (*Meloidogyne incognita*). Open Life Sci 2019; 14(1): 141-9.
[http://dx.doi.org/10.1515/biol-2019-0016] [PMID: 33817146]

[63]   Khanna K, Sharma A, Ohri P, *et al.* Impact of plant growth promoting rhizobacteria in the orchestration of *Lycopersicon esculentum* Mill. resistance to plant parasitic nematodes: a metabolomic approach to evaluate defense responses under field conditions. Biomolecules 2019; 9(11): 676.
[http://dx.doi.org/10.3390/biom9110676] [PMID: 31683675]

[64]   Udalova ZV, Zinovieva SV. Effect of salicylic acid on the oxidative and photosynthetic processes in tomato plants at invasion with root-knot nematode *Meloidogyne incognita* (Kofoid Et White, 1919) Chitwood, 1949. Dokl Biochem Biophys 2019; 488(1): 350-3.
[http://dx.doi.org/10.1134/S160767291905017X] [PMID: 31768858]

[65]   Kayani MZ, Mukhtar T, Hussain MA. Interaction between nematode inoculum density and plant age on growth and yield of cucumber and reproduction of *Meloidogyne incognita.* Pak J Zool 2018; 50(3): 897-902.
[http://dx.doi.org/10.17582/journal.pjz/2018.50.3.897.902]

[66]   Kayani MZ, Mukhtar T, Hussain MA, Ul-Haque MI. Infestation assessment of root-knot nematodes (*Meloidogyne* spp.) associated with cucumber in the Pothowar region of Pakistan. Crop Prot 2013; 47: 49-54.
[http://dx.doi.org/10.1016/j.cropro.2013.01.005]

[67]   Bali S, Kaur P, Sharma A, *et al.* Jasmonic acid-induced tolerance to root-knot nematodes in tomato plants through altered photosynthetic and antioxidative defense mechanisms. Protoplasma 2018; 255(2): 471-84.
[http://dx.doi.org/10.1007/s00709-017-1160-6] [PMID: 28905119]

[68]   Labudda M, Różańska E, Czarnocka W, Sobczak M, Dzik JM. Systemic changes in photosynthesis and reactive oxygen species homeostasis in shoots of *Arabidopsis thaliana* infected with the beet cyst nematode *Heterodera schachtii.* Mol Plant Pathol 2018; 19(7): 1690-704.
[http://dx.doi.org/10.1111/mpp.12652] [PMID: 29240311]

[69]   Cetintas R, Kusek M, Fateh SA. Effect of some plant growth-promoting rhizobacteria strains on root-knot nematode, *Meloidogyne incognita*, on tomatoes. Egypt J Biol Pest Control 2018; 28(1): 7.
[http://dx.doi.org/10.1186/s41938-017-0008-x]

[70]   Sharma IP, Sharma AK. Physiological and biochemical changes in tomato cultivar PT-3 with dual inoculation of mycorrhiza and PGPR against root-knot nematode. Symbiosis 2017; 71(3): 175-83.
[http://dx.doi.org/10.1007/s13199-016-0423-x]

[71]   Mahalik JK, Sahoo NK. Effect of inoculum density of root knot nematode (*Meloidogyne incognita*) on okra (*Abelmoschus esculentus* L.). Int J Plant Prot 2016; 9(2): 603-7.
[http://dx.doi.org/10.15740/HAS/IJPP/9.2/603-607]

[72]   Hussain MA, Mukhtar T, Kayani MZ. Reproduction of *Meloidogyne incognita* on resistant and susceptible okra cultivars. Pak J Agric Sci 2016; 53(2): 371-5.
[http://dx.doi.org/10.21162/PAKJAS/16.4175]

[73]   Lee YS, Kim KY. Antagonistic potential of *Bacillus pumilus* L1 against root-Knot nematode, *Meloidogyne arenaria*. J Phytopathol 2016; 164(1): 29-39.
[http://dx.doi.org/10.1111/jph.12421]

[74]   Lu XH, Sun DQ, Wu QS, Liu SH, Sun GM. Physico-chemical properties, antioxidant activity and mineral contents of pineapple genotypes grown in China. Molecules 2014; 19(6): 8518-32.
[http://dx.doi.org/10.3390/molecules19068518] [PMID: 24959679]

[75]   Azhagumurugan C, Rajan MK. Efficacy of root knot nematode (*Meloidogyne incognita*) on the growth characteristics of black gram (*Vigna mungo*) treated with leaf extract of Magilam (*Mimusops elengi*). Am J Sci Res 2014; 9: 175-81.

[76]   Onyeke CC, Akueshi CO. Infectivity and reproduction of *Meloidogyne incognita* (Kofoid and White) Chitwood on African yam bean, *Sphenostylis stenocarpa* (Hochst Ex. A. Rich) Harms accessions as influenced by botanical soil amendments. Afr J Biotechnol 2012; 11(67): 13095-103.

[77]   Strajnar P, Širca S, Urek G, Šircelj H, Železnik P, Vodnik D. Effect of *Meloidogyne ethiopica* parasitism on water management and physiological stress in tomato. Eur J Plant Pathol 2012; 132(1): 49-57.
[http://dx.doi.org/10.1007/s10658-011-9847-6]

[78]   Kavitha PG, Jonathan EI, Nakkeeran S. Life cycle, histopathology and yield loss caused by root knot nematode, *Meloidogyne incognita* on noni. Madras Agric J 2011; 98(10-12): 386-9.

[79]   Hisamuddin SS, Azam T. Pathogenicity of root-knot nematode, *Meloidogyne incognita* on *Lens culinaris* (Medik.). Arch Phytopathol Pflanzenschutz 2010; 43(15): 1504-11.
[http://dx.doi.org/10.1080/03235400802583537]

[80]   Swain SN, Mahalik JK, Routray BN. Changes in growth traits in blackgram mutant lines induced by different mutagenic treatments towards root knot nematode infection. Assam University J Sci Technol 2010; 4(1): 56-60.

[81]   Ahmed N, Abbasi MW, Shaukat SS, Zaki MJ. Physiological changes in leaves of mungbean plants infected with *Meloidogyne javanica*. Phytopathol medit 2009; 48(2): 262-8.

[82]   Bhat MY, Fazal M, Hissamuddin.Effect of *Meloidogyne incognita* race-1 on the functioning of rhizobial nodules on black gram, *Vigna mungo*. Indian J Nematol 2009; 39(1): 59-64.

[83]   Kimenju JW, Kagundu AM, Nderitu JH, Mambala F, Mutua GK, Kariuki GM. Incorporation of green manure plants into bean cropping systems contribute to root-knot nematode suppression. Asian J Plant Sci 2008; 7(4): 404-8.
[http://dx.doi.org/10.3923/ajps.2008.404.408]

CHAPTER 3

Arsenal Role of Phytochemicals in the Defense System of Plants and the Modulation of Biosynthesis of Phytochemicals

Kapil Paul[1,*]

[1] *Department of Zoology, Kanya Maha Vidyalaya, Jalandhar, Punjab, India*

Abstract: Plants are present ubiquitously on Earth as faunal diversity. Both interact with each other at one or another stage. This interaction can be positive or negative for plants. Interaction for the purpose of pollination is classified as positive interaction whereas faunal diversity (mostly arthropods) is attacking plants to fulfill their food requirement. To defend themselves against this attack by herbivorous animals, plants synthesize some bioactive compounds. Plants release these compounds either to kill or repel these herbivorous animals. Hence it is a direct approach to counter these attacking animals. An indirect approach is also used by some plants where plants produce nectar to attract ants. These ants feed on this nutritious nectar and defend the plants from herbivorous insects that eat the plant's leaves. Compounds synthesized by plants can have noxious odors, and excessive stimulation and some compounds become toxic after ingestion. Ingestion of these compounds can cause many problems such as vomiting, nausea hallucinations convulsions, and even death of the organism.

Keywords: Alkaloid compounds, Plant defense system, Phytochemicals, Phenolics, Terpenoids.

INTRODUCTION

Plants develop complex defense mechanisms against biotic and abiotic stressors when sufficiently opposing forces arise from natural systems. Every living plant cell has the capacity to recognize incoming pathogens and mount an inducible defense against them, which can include the release of poisonous compounds, enzymes that break down infections, and intentional cell death. The production and maintenance of defense-related proteins and poisonous chemicals demand significant energy costs and food requirements, plants frequently wait until infections are discovered before creating these compounds [1]. A large array of

---

* **Corresponding author Kapil Paul:** Department of Zoology, Kanya Maha Vidyalaya, Jalandhar, Punjab, India;
E-mail: kapilpaul09@gmail.com

Shivam Jasrotia & Ajay Kumar (Eds.)

defense mechanisms to fend off disease attacks from fungi, bacteria, and viruses as well as physical, chemical, and biological stressors include cold, drought, heavy metals, and pollution. The release of secondary metabolites like phytoalexins, tannins, and polyphenolic compounds, and the generation of pathogenesis-related (PR) proteins provide this resistance [2]. Around 17 families of such defense related characteristics, such as antibacterial, antifungal, antiviral, anti-oxidative activity, and chitinase and proteinase inhibitory activities, have been found and isolated [3 - 6].

PLANT DEFENSE SYSTEM

A wide range of adversaries, such as infections and herbivores, frequently attack plants. Plants produce phytochemicals as a means of signaling molecules that can deter herbivores and guard against diseases [7, 8]. These phytochemicals are estimated to be more than 2,00,000 low molecular weight which are evolved in response to ecological stressors, namely biotic stressors like herbivore attacks [9, 10]. Most of the plants are reported to produce a variety of phytochemicals, including flavonoids, phenolics, alkaloids, and essential oils [7, 11]. All these substances have the ability to tackle herbivore attacks; some can kill insects instantly upon incorporation, while others can delay or disrupt the development of herbivores, reduce digestive efficiency, thereby lowering resistance to disease and limiting fecundity, repel herbivores, or draw in organisms from a different trophic level [10, 12]. These categories include a number of molecules that protect plants as discussed ahead in Fig. (1).

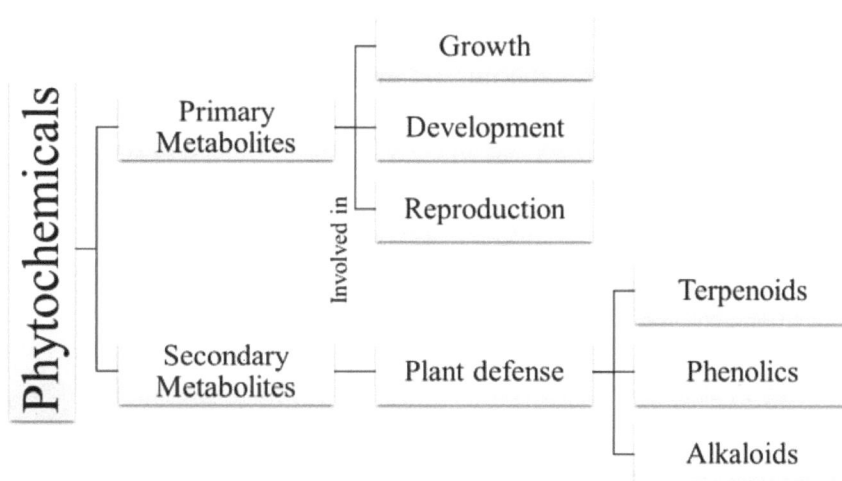

Fig. (1). Functions classification of phytochemicals.

- Isoprene-derived terpenoids including steroids and saponins,
- N-containing alkaloids; (poly)phenolic compounds including flavonoids, tannins; glucosinolates; cyanogenic glucosides,
- Amino acid derivatives such as γ-amino butyric acid (GABA),
- But also peptides/ proteins (proteinase inhibitors, lectins, sporamin); latex; and
- Inorganic compounds (SiO_2, oxalate, selenium) are efficient defensive substances [10, 12].

Terpenoids

Terpenoids play a vital role in inducing chemical deterrents to herbivores. In many plant-pathogen interactions, one of the defense mechanisms against an attack is the synthesis of terpenes, which function as specific or broad pathogen inhibitors. The role that terpenes and terpenoids play in resistance to plant diseases, including bacteria, viruses, and fungi, as well as, when applicable, their vectors are also inhibited [13]. With over 22,000 chemicals reported, terpenoids, or terpenes, are the biggest class of secondary metabolites found in all plants. The most basic terpenoid is the hydrocarbon isoprene (C_5H_8), a volatile gas that is released in enormous quantities during photosynthesis and which may shield cell membranes from harm brought on by intense light or heat. The quantity of isoprene units utilised to create terpenoids determines their classification. Sesquiterpenoids (three units), diterpenoids (four units), and triterpenoids (six units) are among the constituents of monoterpenoids, for instance.

β-ocimene: Ocimenes (3,7-Dimethylocta-1,3,6-triene) are a group of isomeric hydrocarbons. The ocimenes are monoterpenes found within a variety of plants and fruits. The ocimenes are often found naturally as mixtures of the various forms. The mixture, as well as the pure compounds, are oils with a pleasant odor. They are used in perfumery for their sweet herbal scent and are believed to act as plant defense and have anti-fungal properties. In tomatoes and tobacco, β-ocimene defends these plants against pests, particularly *Macrosiphum euphorbiae* [14].

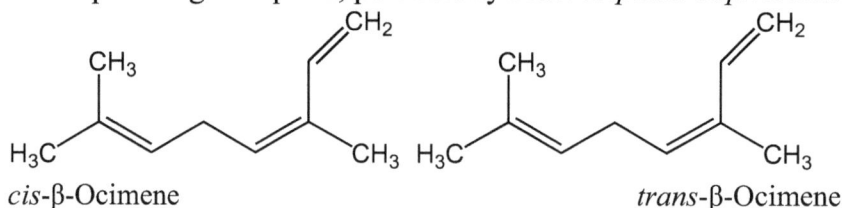

cis-β-Ocimene   *trans*-β-Ocimene

Terpinolene: Terpinolene (3,7-Dimethylocta-1,3,6-triene) is a natural product thathas been isolated from a variety of plant sources like *Camellia sinensis, Hypericum foliosum, Melaleuca alternifolia, etc.* Terpinolene is a p-menthadiene with double bonds at positions 1 and 4 [8]. It has a role as a sedative, an insect repellent, a plant metabolite, and a volatile oil component. They are all colorless liquids with a turpentine-like odor. Terpinolene minimizes the damage caused by the adults of pests (*Paropsisterna tigrina*).

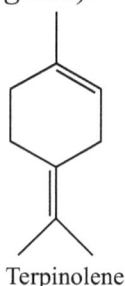

Terpinolene

β-trans-ocimene, (+) limonene: These are colorless liquid aliphatic hydrocarbons classified as cyclic monoterpene, and are the major components in the volatile oil of citrus fruit peels and other plants. The (+) isomer, occurs more commonly in nature as the fragrance of these plants. According to Li *et al.*, (2019) [15], β-tran--ocimene and (+) R-limonene exhibit aphid-eradication properties in the lavender Plant (*Lavandula angustifolia*) in Y-tube olfactometer tests [16].

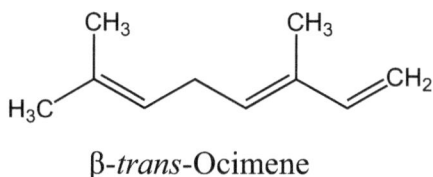

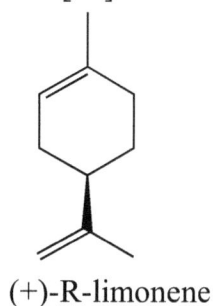

β-*trans*-Ocimene   (+)-R-limonene

S-Methylmethionine: *S*-Methylmethionine $[(CH_3)_2S^+CH_2CH_2CH(NH_3^+)CO_2^-]$ is a derivative of methionine. It deters the wood-boring longicorn beetles belonging to the family Cerambycidae. S-methyl methionine, isolated from a *C. stellata* wood sample, seems to be the main sulfur component contributing to the characteristic odor that is a deterrent to specialist cerambycids seeking oviposition sites [17].

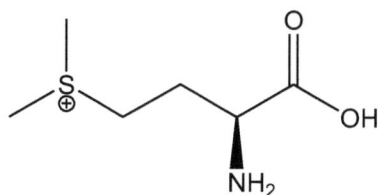

S-Methylmethionine

Tetracyclic terpenes (Cucurbitacins): Cucurbitacins are a class of biological chemicals that some plants, particularly members of the pumpkin and gourd family, Cucurbitaceae, create and operate as a defense against herbivores. Apart from the aforementioned plants, other families (Brassicaceae, Cucurbitaceae, Scrophulariaceae, Begoniaceae, Elaeocarpaceae, Datiscaceae, Desfontainiaceae, Polemoniaceae, Rubiaceae, Sterculiaceae, Rosaceae, and Thymelaeaceae) in certain mushrooms (Russula and Hebeloma), cucurbitacins and their derivatives have also been detected. These compounds have an antibiotic effect on spider mites (*Tetranychus urticae*) but show attractive effects on the pest cucumber beetle.

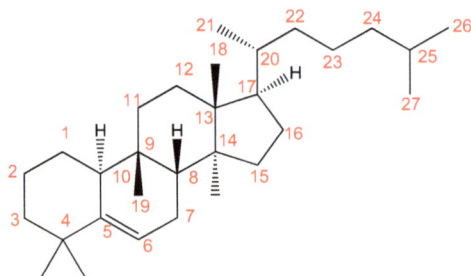

Cucurbitacins

Eugenol, caryophyllene oxide, α -pinene, α -humulene, and α-phellandrene: Eugenol belongs to the allylbenzene class of chemical compounds and is an allyl chain substituted guaiacol. It is an aromatic, oily liquid that ranges in colour from colorless to pale yellow, and it is made up of several essential oils, particularly those from clove, nutmeg, cinnamon, basil, and bay leaf. It is found in clove leaf oil in 82-88% concentrations and in clove bud oil at 80-90%. The aroma of eugenol is spicy, pleasant, and clove-like. *Eugenia caryophyllata*, the previous Linnean nomenclature term for cloves, is the source of the name. *Syzygium aromaticum* is the name that is currently recognized. α-Pinene is an organic compound of the terpene class. It is one of the two isomers of pinene, the other being β-pinene [18]. As an alkene, it contains a reactive four-membered ring. It is found in the oils of many species of coniferous trees, notably the *Pinus* and *Picea* species. It is also found in the essential oil of rosemary (*Rosmarinus officinalis*) and *Satureja myrtifolia* (also known as Zoufa in some regions) [19]. Humulene is one of the components of the essential oil from the flowering cone of the hops plant, *Humulus lupulus*. Phellandrenes are a pair of organic compounds that have a similar molecular structure and similar chemical properties. α -Phellandrene was named after *Eucalyptus phellandra*, now called Eucalyptus *radiata*, from which it can be isolated. It is also a constituent of the essential oil of Eucalyptus dives. β-Phellandrene has been isolated from the oil of water fennel and Canada balsam oil [20]. These compounds are majorly released by Cinnamon and Clove plants and have toxic and repellant effects on adult pests of *Sitophilus granaries* [21].

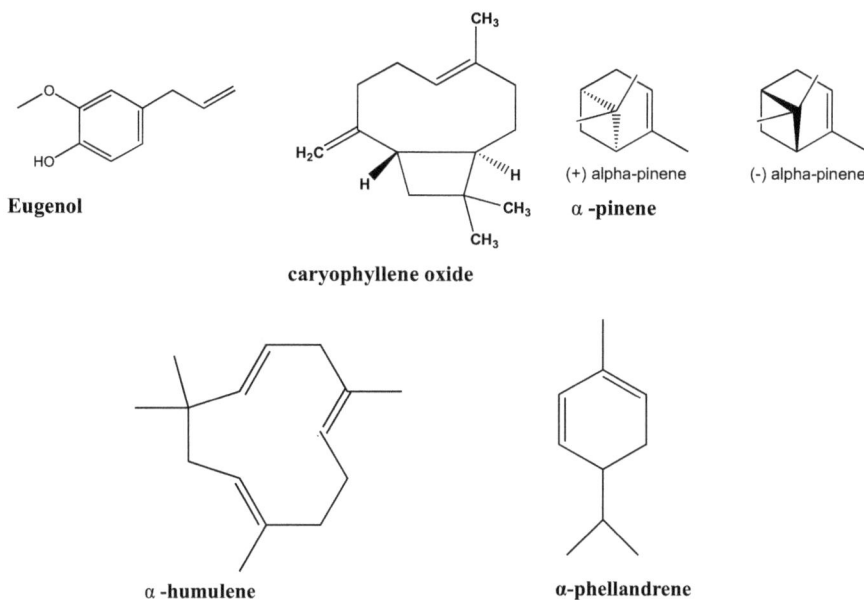

Eugenol

caryophyllene oxide

(+) alpha-pinene   (-) alpha-pinene

α -pinene

α -humulene

α-phellandrene

Phenolics

Phenolic compounds are one of the most predominant and stable secondary metabolites in the plant kingdom. These compounds are known to have diverse roles in plant's physiological activities like nutrition, growth, fertility, and protection from various organisms [22]. Over 8000 such compounds are recognised so far as having immunogenic properties in plants including simple phenolic acids and highly polymerized tannins [23, 24]. Various factors including plant physiology, age, growth stage, environment, and pathogen attack type, all influence the synthesis and concentration of accumulated phenolics [22, 25]. Activities of various phenolic compounds in different plants are discussed ahead.

Grasses and cloves produce a wide range of defense chemicals. *Dipteryx odorata*is known to produce C_6-C_3-Coumarins in order to prevent predation by herbivores [26]. C_6-C_4-Naphthoquinones, such as 2-hydroxynapthoquinone and naphthazarin, have insecticidal action against tobacco culture insects derived from *Calceolaria andina*. 1,4-naphthoquinone derivatives have antimicrobial activity [27, 28]. C_2-C_6-Stilbenes such as trans-resveratrol, trans-piceid, pinosylvin, piceatannol, pinosylvin, trans-pterostilbene, astringin, and rhapontin formed *via* hydroxylation of stilbenoids, operate as phytoalexins in plants [29]. C_6-C_1- C_6-Xanthones are found in the Bonnetiaceae and Clusiaceae families. They are commonly used as an insecticide and ovicide for codling moth eggs [30]. $(C_6$-$C_3)_n$-Lignin and tannins are polymers composed of phenolic chemicals. Lignin is the most abundant source of phenolic substance in plant cell walls.

Ellagitannins, condensed tannins, and gallotannins are various forms of Tannins that impair herbivores' ability to digest plants. 2-Biflavonoids (C_6-C_3-C_6) are created by oxidising two chalcone units and then modifying the middle C_3 units. They are found in gymnosperms such as the Psilotales and Selaginallales, as well as several flowering plants. They are not present in Pinaceae or Gnetales. These bioflavonoids function as fungi toxins and insect-feeding inhibitors [31]. Increased concentration of some phenolic compounds as a response to tissue damage by pathogenic infection and insects [32, 33]. Derivatives of cinnamic acid like p-coumaric acid vanillic acid, and syringic acid are found in rice and cotton plants in order to provide protection from pests [34 - 36]. Increased quantities of phenolic as well as alkaloid compounds induce resistance in plants. Variations in leaf phenolics and alkaloids were observed in Coffea spp. afflicted with *Leucoptera coffeella*, a leaf miner [37]. Phenols in a variety of herbs and spice plant extracts have antibacterial properties that combat infections found in food. It is also known that grape resveratrol has antimicrobial properties [38]. It is well known that fruit and vegetable flavones, flavanones, and phenolic acids are effective against bacteria and fungus *Aspergillus* species (*B. cinerea*, and *F. oxysporum*) [39]. Finnish berries have been shown to possess phenolic chemicals that have antimicrobial effects against probiotic bacteria and other gut bacteria. Lactic acid bacteria and primarily Gram-negative bacteria were reduced in growth by myricetin [40].

Alkaloids

Alkaloids are among the most abundant classes of secondary metabolites and are primarily involved in defensive purposes in plants [41]. These are nitrogens carrying at least a nitrogen group in a heterocyclic ring and are thought to have isoprenoids, several amino acids, and purine nucleotides as their precursors [42, 43]. Their biosynthesis typically starts in the roots, and then moves *via* the phloem and typically the xylem [44]. Earlier, alkaloids were categorized based on the genera of plants in which they were found. More than 2000 alkaloids and 10-15% are produced by plants belonging to about 40 different plant families including Amaryllidaceae, Solanaceae, Liliaceae, and Leguminosae [45, 46]. Several studies have demonstrated many alkaloids like benzylisoquinoline alkaloids and their derivatives have anti-herbivory, antifungal, and/or antibacterial qualities in the plant kingdom enlisted in Table **1**.

Table 1. List of various alkaloids, activities, and their structures.

Compound	Activity	Structure	References
Tabienine B	Insecticidal Larvicidal		[47]
Hunnemanine	Antifungal		[48]
Hydrastine	Antifungal		[49, 50]
Chelerythrine	Antifingal		[50]
Sanguinarine	Antifungal Antibacterial Antiviral		[51, 52]
Anolobine	Antifungal Antibacterial		[53]
Liriodenine	Antifungal Antibacterial		[53, 54]

(Table 1) cont.....

Compound	Activity	Structure	References
Papaverine	Feeding deterrent		[55]
Berberine	Antifungal Antibacterial Antiviral Feeding Deterrent		[50, 51, 55 - 58]

All these compounds (phenolics, terpenoids, and alkaloids) are biosynthesized, stored, and released under attack situations. The defense system of plants includes chemicals, protective proteins, enzymes, and resin deposits that can flow to repel or physically trap small insects for their protection [59, 60].

CONCLUSION

Plants have developed chemical and structural defenses to withstand a variety of environmental stresses. Each and every live plant cell possesses the ability to detect intruding pathogens and initiate an inducible defense against them. This defense may take the form of noxious substances being released, infection-breaking enzymes, or deliberate cell death. Significant energy and food needs are required for the synthesis and maintenance of defense-related proteins and toxic chemicals; plants typically delay producing these substances until diseases are. A wide range of plant defense systems are there to prevent bacterial, viral, and fungal disease attacks. These chemical compounds are synthesized and released at the time of any kind of chemical and biological attack. Hence these plants play a vital role in the prevention of herbivorous insects at various stages.

REFERENCES

[1]   Freeman BC, Beattie GA. An overview of plant defenses against pathogens and herbivores. Plant Health Instructor 2008; 8(1): 1-12.

[2]   Tam JP, Wang S, Wong KH, Tan WL. Antimicrobial peptides from plants. Pharmaceuticals (Basel) 2015; 8(4): 711-57.
       [http://dx.doi.org/10.3390/ph8040711] [PMID: 26580629]

[3]   Ebrahim S, Kalidindi U, Singh B. Pathogenesis related (PR) proteins in plant defense mechanism. Science against Microbial pathology. 2011; 2(3): 1043–54. Available from: https://www.researchgate.net/publication/284957170

[4]   Sels J, Mathys J, De Coninck BMA, Cammue BPA, De Bolle MFC. Plant pathogenesis-related (PR) proteins: A focus on PR peptides. Plant Physiol Biochem 2008; 46(11): 941-50.
       [http://dx.doi.org/10.1016/j.plaphy.2008.06.011] [PMID: 18674922]

[5]   Sinha M, Singh RP, Kushwaha GS, *et al.* Current overview of allergens of plant pathogenesis related protein families. Sci World J 2014.

[6]   Stintzi A, Heitz T, Prasad V, *et al.* Plant 'pathogenesis-related' proteins and their role in defense against pathogens. Biochimie 1993; 75(8): 687-706.
[http://dx.doi.org/10.1016/0300-9084(93)90100-7] [PMID: 8286442]

[7]   Soliman SSM, Saeed BQ, Elseginy SA, *et al.* Critical discovery and synthesis of novel antibacterial and resistance-modifying agents inspired by plant phytochemical defense mechanisms. Chem Biol Interact 2021; 333(1): 109318.
[http://dx.doi.org/10.1016/j.cbi.2020.109318] [PMID: 33186599]

[8]   Kokoska L, Havlik J, Valterova I, Sovova H, Sajfrtova M, Jankovska I. Comparison of chemical composition and antibacterial activity of *Nigella sativa* seed essential oils obtained by different extraction methods. J Food Prot 2008; 71(12): 2475-80.
[http://dx.doi.org/10.4315/0362-028X-71.12.2475] [PMID: 19244901]

[9]   Pichersky E, Lewinsohn E. Convergent evolution in plant specialized metabolism. Annu Rev Plant Biol 2011; 62(1): 549-66.
[http://dx.doi.org/10.1146/annurev-arplant-042110-103814] [PMID: 21275647]

[10]   Mithöfer A, Maffei ME. General mechanisms of plant defense and plant toxins. Gopalakrishnakone P, Carlini CR, Ligabue-Braun R Eds, Plant toxins. 2017. p. 3–24.
[http://dx.doi.org/10.1007/978-94-007-6464-4_21]

[11]   Badria FA, Ameen M, Akl MR. Evaluation of cytotoxic compounds from *Calligonum comosum* L. Growing in Egypt. Z Naturforsch C J Biosci. 2007 ;62(9-10): 656-60. Available from: http://www.znaturforsch.com

[12]   War AR, Paulraj MG, Ahmad T, *et al.* Mechanisms of plant defense against insect herbivores. Plant Signal Behav. 2012; 7(10): 1306-20.

[13]   Toffolatti SL, Maddalena G, Passera A, Casati P, Bianco PA, Quaglino F. Role of terpenes in plant defense to biotic stress. In: Jogaiah S Ed, Biocontrol agents and secondary metabolites: applications and immunization for plant growth and protection. Woodhead Publishing 2021; pp. 401-17.
[http://dx.doi.org/10.1016/B978-0-12-822919-4.00016-8]

[14]   Cascone P, Iodice L, Maffei ME, Bossi S, Arimura G, Guerrieri E. Tobacco overexpressing β-ocimene induces direct and indirect responses against aphids in receiver tomato plants. J Plant Physiol 2015; 173: 28-32.
[http://dx.doi.org/10.1016/j.jplph.2014.08.011] [PMID: 25462075]

[15]   Li H, Li J, Dong Y, *et al.* Time-series transcriptome provides insights into the gene regulation network involved in the volatile terpenoid metabolism during the flower development of lavender. BMC Plant Biol 2019; 19(1): 313.
[http://dx.doi.org/10.1186/s12870-019-1908-6] [PMID: 31307374]

[16]   Boncan DAT, Tsang SSK, Li C, Lee IHT, Lam HM, Chan TF, *et al.* Terpenes and terpenoids in plants: Interactions with environment and insects. Int J Mol Sci. 2020 Oct 6; 21(19): 7382.

[17]   Berkov A, Meurer-Grimes B, Purzycki KL. Do Lecythidaceae Specialists (Coleoptera, Cerambycidae) Shun Fetid Tree Species? Biotropica 2000; 32(3): 440-51.
[http://dx.doi.org/10.1111/j.1744-7429.2000.tb00491.x]

[18]   Simonsen JL. The Terpenes. 2nd ed. Vol. 2. London, United Kingdom: Cambridge University Press; 1957.   105–191.   Available   from: https://discovered.ed.ac.uk/permalink/44UOE_INST/iatqhp/alma99824233502466

[19]   Zebib B, Beyrouthy MEL, Safi C, Merah O. Chemical composition of the essential oil of *Satureja myrtifolia* (Boiss. & Hohen.) from Lebanon. J Essent Oil-Bear Plants 2015; 18(1): 248-54.
[http://dx.doi.org/10.1080/0972060X.2014.890075]

[20]   Boland DJ, Brophy JJ, House PN. Eucalyptus leaf oils, use, chemistry, distillation and marketing. In 1991.
[http://dx.doi.org/10.1002/ffj.2730070209]

[21]   Plata-Rueda A, Campos JM, da Silva Rolim G, *et al.* Terpenoid constituents of cinnamon and clove essential oils cause toxic effects and behavior repellency response on granary weevil, *Sitophilus granarius*. Ecotoxicol Environ Saf 2018; 156: 263-70.
[http://dx.doi.org/10.1016/j.ecoenv.2018.03.033] [PMID: 29554611]

[22]   Pratyusha S. Phenolic compounds in the plant development and defense: an overview. In: Hasanuzzaman M, Nahar K, editors. Plant stress physiology - perspectives in agriculture. 2022. Available from: www.intechopen.com

[23]   Beninger CW, Abou-Zaid MM, Kistner ALE, *et al.* A flavanone and two phenolic acids from *Chrysanthemum morifolium* with phytotoxic and insect growth regulating activity. J Chem Ecol 2004; 30(3): 589-606.
[http://dx.doi.org/10.1023/B:JOEC.0000018631.67394.e5] [PMID: 15139310]

[24]   Harborne JB. General procedures and measurement of total phenolics. In: Dey PM, Harborne JB, Eds. Methods in plant biochemistry. London: Academic Press 1989; pp. 1-28.
[http://dx.doi.org/10.1016/B978-0-12-461011-8.50007-X]

[25]   Ozyigit II, Kahraman V, Ercan O. Relation between explant age, total phenols and regeneration response in tissue cultured cotton (*Gossypium hirsutum* L.). Afr J Biotechnol 2007; 6(1): 3-008. Available from: http://www.academicjournals.org/AJB

[26]   Sarker SD, Nahar L. Progress in the chemistry of naturally occurring coumarins. In: Kinghorn AD, Falk H, Gibbons S, Kobayashi J, Eds. Progress in the chemistry of organic natural products 106. Cham: Springer International Publishing 2017; pp. 241-304.
[http://dx.doi.org/10.1007/978-3-319-59542-9_3]

[27]   Ribeiro KAL, de Carvalho CM, Molina MT, *et al.* Activities of naphthoquinones against Aedes aegypti (Linnaeus, 1762) (Diptera: Culicidae), vector of dengue and *Biomphalaria glabrata* (Say, 1818), intermediate host of *Schistosoma mansoni*. Acta Trop 2009; 111(1): 44-50.
[http://dx.doi.org/10.1016/j.actatropica.2009.02.008] [PMID: 19426662]

[28]   Khambay BPS, Jewess P. The potential of natural naphthoquinones as the basis for a new class of pest control agents — an overview of research at IACR-Rothamsted. Crop Prot 2000; 19(8-10): 597-601. Available from: https://www.sciencedirect.com/science/article/pii/S0261219400000788
[http://dx.doi.org/10.1016/S0261-2194(00)00078-8]

[29]   Chong J, Poutaraud A, Hugueney P. Metabolism and roles of stilbenes in plants. Plant Sci 2009; 177(3):   143-55.   Available   from: https://www.sciencedirect.com/science/article/pii/S0168945209001551
[http://dx.doi.org/10.1016/j.plantsci.2009.05.012]

[30]   Monajjemi M, Azizi V, Amini S, Mollaamin F. Nanotheoretical studies on evaluation of anti cancer potential on mangosteen plant. Afr J Agric Res 2011; 6(19): 4661-70. Available from: http://www.academicjournals.org/AJAR

[31]   Iwashina T. Flavonoid function and activity to plants and other organisms. Biol Sci Space 2003; 17(1): 24-44.
[http://dx.doi.org/10.2187/bss.17.24] [PMID: 12897458]

[32]   Khemani LD, Srivastava MM, Srivastava S, Eds. Rehman. F, Khan. F. A., Badruddin SMA. Role of phenolics in plant defense against insect herbivory. Chemistry of phytopotentials: health, energy and environmental perspectives. Heidelberg, Berlin: Springer 2012.
[http://dx.doi.org/10.1007/978–3-642–23394-4_65]

[33]   Johnson KS, Felton GW. Plant phenolics as dietary antioxidants for herbivorous insects: a test with genetically modified tobacco. J Chem Ecol 2001; 27(12): 2579-97.

[http://dx.doi.org/10.1023/A:1013691802028] [PMID: 11789960]

[34] Pathipati UR, Yasur J. Physiological changes in groundnut plants induced by pathogenic infection of cercosporidium personatum deighton. Allelopathy J 2009; 23(2): 369-78.

[35] Hildebrand DF, Rodriguez JG, Brown GC, Luu KT, Volden CS. Peroxidative responses of leaves in two soybean genotypes injured by twospotted spider mites (Acari: Tetranychidae). J Econ Entomol 1986; 79(6): 1459-65.
[http://dx.doi.org/10.1093/jee/79.6.1459]

[36] Usha Rani P, Pratyusha S. Defensive role of *Gossypium hirsutum* L. anti-oxidative enzymes and phenolic acids in response to *Spodoptera litura* F. feeding. J Asia Pac Entomol 2013; 16(2): 131-6.
[http://dx.doi.org/10.1016/j.aspen.2013.01.001]

[37] Bi JL, Murphy JB, Felton GW. Antinutritive and oxidative components as mechanisms of induced resistance in cotton to helicoverpa zea. J Chem Ecol 1997; 23(1): 97-117.
[http://dx.doi.org/10.1023/B:JOEC.0000006348.62578.fd]

[38] Ma DSL, Tan LTH, Chan KG, *et al.* Resveratrol-potential antibacterial agent against foodborne pathogens. Front Pharmacol 2018; 9(2): 102.
[http://dx.doi.org/10.3389/fphar.2018.00102] [PMID: 29515440]

[39] Bartmańska A, Wałecka-Zacharska E, Tronina T, *et al.* Antimicrobial properties of spent hops extracts, flavonoids isolated therefrom, and their derivatives. Molecules 2018; 23(8): 2059.
[http://dx.doi.org/10.3390/molecules23082059] [PMID: 30126093]

[40] Puupponen-Pimia R, Nohynek L, Meier C, Ka M, Hko È Nen È, Heinonen M, *et al.* Antimicrobial properties of phenolic compounds from berries. J Appl Microbiol. 2001; 90(4): 494-507.

[41] Kant MR, Jonckheere W, Knegt B, *et al.* Mechanisms and ecological consequences of plant defence induction and suppression in herbivore communities. 2015
[http://dx.doi.org/10.1093/aob/mcv054]

[42] Itkin M, Heinig U, Tzfadia O, Bhide AJ, Shinde B, Cardenas PD, *et al*. Biosynthesis of antinutritional alkaloids in solanaceous crops is mediated by clustered genes. Science. 2013; 341(6142): 175–9.

[43] Levin DA. Alkaloid-bearing plants: an ecogeographic perspective. Am Nat 1976; 110(972): 261-84. Available from: http://www.jstor.org/stable/2459491
[http://dx.doi.org/10.1086/283063]

[44] Courdavault V, Papon N, Clastre M, Giglioli-Guivarc'h N, St-Pierre B, Burlat V. A look inside an alkaloid multisite plant: the Catharanthus logistics. Curr Opin Plant Biol 2014; 19: 43-50.
[http://dx.doi.org/10.1016/j.pbi.2014.03.010] [PMID: 24727073]

[45] Howe GA, Jander G. Plant immunity to insect herbivores. Annu Rev Plant Biol 2008; 59(1): 41-66.
[http://dx.doi.org/10.1146/annurev.arplant.59.032607.092825] [PMID: 18031220]

[46] Fürstenberg-Hägg J, Zagrobelny M, Bak S. Plant defense against insect herbivores. Int J Mol Sci 2013; 14(5): 10242-97.
[http://dx.doi.org/10.3390/ijms140510242] [PMID: 23681010]

[47] Quevedo R, Núñez L, Moreno B. A rare head–head binding pattern in bisbenzylisoquinoline alkaloids. Nat Prod Res 2011; 25(9): 934-8.
[http://dx.doi.org/10.1080/14786419.2010.512000] [PMID: 21547845]

[48] Singh S, Jain L, Pandey MB, Singh UP, Pandey VB. Antifungal activity of the alkaloids from *Eschscholtzia californica*. Folia Microbiol (Praha) 2009; 54(3): 204-6.
[http://dx.doi.org/10.1007/s12223-009-0032-7] [PMID: 19649736]

[49] Goel M, Singh UP, Jha RN, Pandey VB, Pandey MB. Individual and combined effect of (±)- α-hydrastine and (±)-β-hydrastine on spore germination of some fungi. Folia Microbiol (Praha) 2003; 48(3): 363-8.
[http://dx.doi.org/10.1007/BF02931368] [PMID: 12879748]

[50]   Tims MC, Batista C. Effects of root isoquinoline alkaloids from *Hydrastis canadensis* on *Fusarium oxysporum* isolated from Hydrastis root tissue. J Chem Ecol 2007; 33(7): 1449-55.
[http://dx.doi.org/10.1007/s10886-007-9319-9] [PMID: 17549565]

[51]   Schmeller T, Latz-Brüning B, Wink M. Biochemical activities of berberine, palmatine and sanguinarine mediating chemical defence against microorganisms and herbivores. Phytochemistry 1997; 44(2): 257-66.
[http://dx.doi.org/10.1016/S0031-9422(96)00545-6] [PMID: 9004542]

[52]   Liu H, Wang J, Zhao J, *et al.* Isoquinoline alkaloids from *Macleaya cordata* active against plant microbial pathogens. Nat Prod Commun 2009; 4(11): 1934578X0900401120.
[http://dx.doi.org/10.1177/1934578X0900401120] [PMID: 19967990]

[53]   Villar A, Mares M, Rios JL, Canton E, Gobernado M. Antimicrobial activity of benzylisoquinoline alkaloids. Pharmazie 1987; 42(4): 248-50.
[PMID: 3615557]

[54]   Costa EV, Pinheiro MLB, Barison A, *et al.* Alkaloids from the bark of *Guatteria hispida* and their evaluation as antioxidant and antimicrobial agents. J Nat Prod 2010; 73(6): 1180-3.
[http://dx.doi.org/10.1021/np100013r] [PMID: 20476748]

[55]   Sellier MJ, Reeb P, Marion-Poll F. Consumption of bitter alkaloids in *Drosophila melanogaster* in multiple-choice test conditions. Chem Senses 2011; 36(4): 323-34.
[http://dx.doi.org/10.1093/chemse/bjq133] [PMID: 21173029]

[56]   Krug E, Proksch P. Influence of dietary alkaloids on survival and growth of *Spodoptera littoralis*. Biochem Syst Ecol 1993; 21(8): 749-56.
[http://dx.doi.org/10.1016/0305-1978(93)90087-8]

[57]   Park IK, Lee HS, Lee SG, Park JD, Ahn YJ. Antifeeding activity of isoquinoline alkaloids identified in *Coptis japonica* roots against *Hyphantria cunea* (Lepidoptera: Arctiidae) and *Agelastica coerulea* (Coleoptera: Galerucinae). J Econ Entomol 2000; 93(2): 331-5.
[http://dx.doi.org/10.1603/0022-0493-93.2.331] [PMID: 10826181]

[58]   Shields VDC, Smith KP, Arnold NS, Gordon IM, Shaw TE, Waranch D. The effect of varying alkaloid concentrations on the feeding behavior of gypsy moth larvae, *Lymantria dispar* (L.) (Lepidoptera: Lymantriidae). Arthropod-Plant Interact 2008; 2(2): 101-7.
[http://dx.doi.org/10.1007/s11829-008-9035-6] [PMID: 21278814]

[59]   Sánchez HS, Contreras AM. Chemical Plant Defense Against Herbivores. Herbivores. InTech 2017; pp. 3-28.

[60]   Franceschi VR, Nakata PA. CALCIUM OXALATE IN PLANTS: Formation and Function. Annu Rev Plant Biol 2005; 56(1): 41-71.
[http://dx.doi.org/10.1146/annurev.arplant.56.032604.144106] [PMID: 15862089]

CHAPTER 4

Phytochemicals' Classes Involved in Nematode Defense and their Related Activities

Urvashi Dhiman[1,2]**, Prasoon Gupta**[1,2] **and Ripu Daman Parihar**[3,*]

[1] *Natural Products and Medicinal Chemistry Division, CSIR- Indian Institute of Integrative Medicine, Canal Road, Jammu 180001, India*

[2] *Academy of Scientific and Innovative Research, CSIR-Human Resource Development Centre, Ghaziabad 201002, India*

[3] *Department of Zoology, University of Jammu, Jammu & Kashmir 174303, India*

Abstract: Plant parasitic nematodes (PPNs) are one of the most lethal pests that have emerged in the past years. These nematodes are microscopic in size, cylindrical in shape, and inhabit mostly terrestrial ecosystems. They account for a significant biotic limiting factor that hampers crop yield and productivity. PPNs are majorly categorized into three categories such as lesion nematodes (*Pratylenchus* spp.), Root-knot nematodes (*Meloidogyne* spp.), and cyst nematodes (*Heterodera* and *Globodera* spp.). They are known to be the primary cause of pest infestation among other PPNs. Terpenes, flavonoids, alcoholics, and phenolics are essential plant secondary metabolites with a reliable potential to control the PPN population. Reports have shown that they reduce the gall size, inhibit egg hatching, increase the mortality rate of infective juveniles (IJs), *etc.*, which lead to the death of IJs and hence protect the crops against PPNs. Such studies elucidate the importance of using plant phytoconstituents as a natural alternative to hazardous chemical pesticides, which are dangerous to humankind and nature. This chapter culminates the efficiency of plant secondary metabolites and their significance in killing root-knot nematodes majorly of different species infesting commercial agricultural crops at different life cycle stages.

Keywords: Egg hatching, Infective juvenile, Plant parasitic nematodes, Root knot nematode, Secondary metabolites.

INTRODUCTION

Agriculture contributes significantly (4.3%) to the world's Gross Domestic Product (GDP). Plant Parasitic Nematodes (PPNs) are profoundly known to cause severe damage to most of the agriculturally important plants in high demand and

* **Corresponding author Ripu Daman Parihar:** Department of Zoology, University of Jammu, Jammu & Kashmir 174303, India; E-mail: ripuparihar@jammuuniversity.ac.in

Shivam Jasrotia & Ajay Kumar (Eds.)

consumed by people worldwide. Their attack is responsible for 21.3% of crop loss, which is equivalent to nearly 1.58 billion USD yearly, in which 19 horticultural crops were assessed at a loss of 50,224.98 million and 11 field crop losses of 51,814.81 million. Among PPNs, the *Meloidogyne graminicola*, a root-knot nematode for rice, is responsible for a loss in the crop yield of 23,272.32 million [1]. Fruit crops such as citrus and banana suffered losses of 9828.22 and 9710.46 million; other vegetable crops such as tomato and okra bear losses of 6035.2 and 2480.86 million [1]. In recent years, researchers have developed many strategies to combat the action of PPNs to alleviate the burden of loss of the agricultural economy. Plant secondary metabolites have shown remarkable and promising results lately against different species of plant parasitic nematodes at different stages of the nematode life cycle. Herein, we have discussed the plant secondary metabolites belonging to different classes and their action against the plant parasitic nematodes.

PLANT SECONDARY METABOLITES

The medicinal plants show diverse pharmacological activities due to the phytochemical constituents present in them. Based on their involvement in the metabolic processes, they are broadly categorized into primary and secondary metabolites. Primary ones play an important role in the basic metabolic and physiological functioning of life and are found similar in most living cells. The secondary metabolites play a vital role in alleviating many ailments in traditional medicine systems and folk uses. Plant metabolites prove to be a boon in the field of drug discovery in modern medical systems. Secondary plant metabolites are classified according to their chemical structures into various classes such as terpenes, phenolics, alkaloids, flavonoids, glucosinolates, *etc.* [2].

The secondary metabolites show different biological effects that provide a scientific basis for the utility of herbs in many traditional practices used by ancient communities across the globe. They have shown antifungal, antiviral, and antibiotic properties and thus protect the plants from lethal pathogens. The list of phytoconstituents and their action have been provided in Table **1** and chemical structure and target action are presented in Figs. (**1** and **2**).

Table 1. List of phytochemical constituents and their nematicidal activity against different stages and species of plant parasitic nematodes.

Secondary Metabolite Source	Secondary Metabolites and the Class of Phytoconstituent	Mode of Action	Crop	Target Nematode	Inference	References
Artemisia elegantissimia A. incisa	Isoscopletin (Coumarin), Carbofuran and Apigenin (Flavonoid)	Mortality of J2s and egg hatch inhibition	Tomato	*Meloidogyne incognita*	Inhibition with (90.0%) and (96.0%) at 0.3 mg/mL concentration.	[27]

(Table 1) cont.....

Secondary Metabolite Source	Secondary Metabolites and the Class of Phytoconstituent	Mode of Action	Crop	Target Nematode	Inference	References
Carvone, cuminaldhyde, linalool, and cineole	J2 hatching inhibition and mortality.	Inhibition of egg hatching	Tomato	*Meloidogyne incognita*	LC$_{50}$ values: 123.5, 172.2, 354.9, 466.4, & 952.3 µg/mL, respectively.	[31]
Aristolochia mollissima	Aristolochic acid I, aristololactam I, aristololactam W	J2 Mortality	-	*Meloidogyne javanica*	LC$_{50}$ values of 45.25, 36.56, 119.46 mg· L^{-1} after 96 h.	[32]
Syzygium aromaticum	*para* methyl benzoic acid, 3-*O-trans-para*-coumaroylmaslinic acid, methyl maslinate, maslinic acid	-	-	*Meloidogyne incognita*	Mortality inhibition (88–92%	[33]
Lavandula intermedia (Other species: *abrialis, cerioni, sumiens*)	Linalool (Terpene)	Inhibition of egg hatching along with J2 mortality; reduction of galls, eggs	Tomato	*Meloidogyne incognita*	EO; 24.9 µg/mL-1, 1.2 µg/mL-1, 17.4 µg/mL-1	[34]
Lavandula intermedia (Other species: *abrialis, Cerioni, sumiens*)	Linalool (Terpene)	Inhibition of egg hatching along with J2 mortality	-	*Pratylenchus vulnus*	Mortality Rate 65.5%, 67.7%, and 75.7% (4 h of exposure)	[34]
Monarda didyma, Monarda fistulosa	γ-terpinene (Terpenoid), o-cymene(Aromatic Hydrocarbon), Carvacrol(Monoterpenoid)	J2 mortality, egg-hatching inhibition, galls and eggs reduction in soil	Tomato	*Meloidogyne incognita*	(J2 mortality)1.0 µL mL−1; 12.5 µL mL−1 (24 h) (egg hatching)500 &1000 µg mL−1 (24-48 h)	[35]
Mentha longifolia	piperitone oxide (Monoterpene)	Inhibition of J2 mortality and egg hatching inhibition	-	*Meloidogyne graminicola*	EO; 15.62-1000 ppm (96 h)	[36]
Mentha spicata L.	Carvone (Monoterpene), limonene(Cyclic Monoterpene)	J2 mortality, reduction of galls and eggs	Coleus	*Meloidogyne javanica*	EO; 1000, 2000, 3000, 4000, & 5000 ppm (v/v) for 24 h; 48 h; 72 h	[37]
Thymus vulgaris L.	Thymol, ρ-cymene	J2 mortality, Reduction of the galls and eggs	Coleus	*Meloidogyne javanica*	EO; 1000, 2000, 3000, 4000,& 5000 ppm (v/v) (24 h; 48 h;72 h)	[37]
Mentha spicata L.	-	-	Pepper	*Meloidogyne incognita*	EO; 5% (v/v) (72 h)	[38]
Basilicum L.	SabineneMonoterpene), Myrcene(Monoterpene) Transcaryophyllene (Sesquiterpene)	J2 mortality, reduction of galls	Pepper	*Meloidogyne incognita*	EO; 5% (v/v) (72 h)	[38]

(Table 1) cont.....

Secondary Metabolite Source	Secondary Metabolites and the Class of Phytoconstituent	Mode of Action	Crop	Target Nematode	Inference	References
Trifolium incarnatum	(z)-3-hexenyl acetates, (Z)-3-hexane-1-ol, (E) -ocimene, furanoeudesm-1,3-diene	J2 mortality, reduction of the galls	Chili pepper (Capsicum annuum L.)	*Meloidogyne incognita*	EO; 3% and 5% (v/v) (48 h)	[38]
Nepeta cateria	-	J2 mortality	-	*Meloidogyne incognita*	EO; 1.2 mL/L	[39]
Ocimum sanctum L.	Eugenol (Phenolics)methyl ether	J2 mortality	-	*Meloidogyne incognita*	EO; 1230 mg/L (24 h)	[39]
Cinnamomum zeylanicum Blume	Eugenol	J2 mortalityand J2 paralysis	-	*Meloidogyne incognita*	EO; 391 mg/L (24 h)	[39]
Ocimum basilicum	i-linalool (Monoterpene)	J2 mortality	-	*Meloidogyne hapla*	-	[40]
Foeniculum vulgare	Anethole ii-Monoterpene	J2 mortality	-	*Meloidogyne hapla*	-	[40]
Pogostemon cablin Benth	α-guaiene(sesquiterpene), patchoulolsesquiterpene α-bulnesene sesquiterpene	J2 mortality, J2 immobility	-	*Meloidogyne incognita*	EO; 250 μg/mL−1, 31.25 μg/mL−1 (24 h)	[41]
Pogostemon cablin	α-guaiene	Mortality and paralysis	Pepper	*Meloidogyne incognita*	EO; 387.77 μg mL−1 (48 h)	[42]
Artemisia absinthium	β-terpineol	Mortality and paralysis of J2	-	*Meloidogyne incognita*	EO; 937.52 μg mL−1 (48 h)	[42]
Acorus calamus	β-asarone	J2 mortality, J2 paralysis	-	*Meloidogyne incognita*	EO: 524.45 μg mL−1 for 24 h	[42]
Commiphora myrrha	furanoeudesm-1,3-diene, curcerene	J2 mortality and paralysis	-	*Meloidogyne incognita*	EO; 1000 μg mL−1 (24 h)	[42]
Eucalyptus citriodora	Citronellal (**Monoterpene**)	J2 mortality and J2 paralysis	-	*Meloidogyne incognita*	EO; 746.48 μg mL−1 (24 h)	[42]
Melaleuca alternifolia	β-terpineol,γ-terpinene (**Monoterpene**)	J2 mortality and paralysis	-	*Meloidogyne incognita*	EO; 404.13 μg mL−1 (24 h)	[42]
Myrtus communis	α-pinene, 1,8-cineol(**Terpene**)	J2 mortality, J2 paralysis	-	*Meloidogyne incognita*	EO; 932.65 μg mL−1 for 48 h	[42]
Citrus sinensis	l-limonene(**Monoterpene**)	J2 mortality and paralysis	-	*Meloidogyne incognita*	EO; 353.20 μg mL−1 (24 h)	[42]
Thymus citriodorus	Geraniol Monoterpene	Paralysis and mortality of J2 stage	Tomato	*Meloidogyne incognita*	EO; 50 μL kg−1 soil	[43]
Teucrium polium	Limonene; α-pinene; β-pinene	J2 mortality	-	*Meloidogyne incognita*	EO; 4000&8000 ppm (v/v) (24 h)	[44]
Achillea wilhelmsii	1,8-cineole, Limonene Monoterpene, α-pinene, β-pinene	J2 mortality	-	*Meloidogyne incognita*	EO; 4000 and 8000 ppm (v/v) (24 h)	[44]
Tanacetum polium	(e)-caryophyllene, limonene, α-pinene, β-pinene	J2 mortality	-	*Meloidogyne incognita*	EO; 4000 & 8000 ppm (v/v) (24 h)	[44]

(Table 1) cont.....

Secondary Metabolite Source	Secondary Metabolites and the Class of Phytoconstituent	Mode of Action	Crop	Target Nematode	Inference	References
Thymus linearis Benth	Thymol, carvacrol	Inhibition of egg hatching and J2 mortality	-	*Meloidogyne incognita*	Rainy season: 5 µL/mL (72 h) (J2 mortality), 2 µL/mL (72 h) (egg hatching) Winter season: 8 µL/m (72 h) (J2 mortality),2 µL/mL (72h) h(egg hatching)	[45]
Dysphania ambrosioides	(z)-ascaridole Monoterpene, e-ascaridole, p-cymene	J2 mortality, egg-hatching inhibition, reduction of galls and eggs	Tomato	*Meloidogyne incognita*	EO: 500 µg mL−1 (48 h)	[46]
Allium sativum	Diallyl disulfide, Diallyl trisulfide, Methyl allyl trisulfide	Inhibition of egg hatching and J2 mortality	Tomato	*Meloidogyne javanica*	EO; 0.025 µg mL−1 (72 h) Hydrolat: 0.125 µg mL−1 (72 h)	[47]
Cuminum cyminum	γ-terpinen-7-al; α-terpinen-7-al; cumin aldehydes	Egg paralysis as well as Inhibition of egg hatching and J2 mortality differentiation, reduction of nematode population in soil	Tomato	*Meloidogyne javanica*	EO; 62.5 µL/L, 2000 µL/L for 48 & 96 h of immersion	[48]
Daucus carota	Carotol sesquiterpene, daucol; daucene	Inhibitioon of egg hatching and J2 mortality	-	*Meloidogyne incognita*	EO; 2500 ppm (96 h)	[49]
Laurus nobilis L.	linalool, 1,8-cineole, α-pinene, β-pinene, α-terpinyl acetate (**Monoterpene**)	J2 mortality, egg-hatching inhibition	-	*Meloidogyne incognita*	EO; 0.80 µg/mL−1 (96 h)	[49]
Ridolfia segetum	(z)-β-ocimene, β-pinene	J2 mortality and mobility; egg-hatching inhibition	-	*Meloidogyne javanica*	EO; 16 µL/mL (72 h)	[50]
Artemisia nilagirica	α-thujone Monoterpene, α-myrcene, linalyl isovalerate, camphor, caryophyllene oxide, eucalyptol	J2 mortality, egg-hatching inhibition, reduction of the galls, eggs, & nematodes in the soil	Tomato	*Meloidogyne incognita*	EO; 20 µg/mL (48 h)	[51]
Tanacetum falconeri	cisdehydromatricaria ester-1	J2 mortality	-	*Meloidogyne incognita*	EO; 1% (w/v) (24 h)	[52]
Brassica nigra	Allyl isothiocyanate (AITC)	J2 mortality and paralysis	-	*Meloidogyne incognita*	EO; 47.7 µg mL−1 (72 h)	[53]

(Table 1) cont.....

Secondary Metabolite Source	Secondary Metabolites and the Class of Phytoconstituent	Mode of Action	Crop	Target Nematode	Inference	References
Piptadenia viridiflora	Benzaldehyde	J2 mortality	-	*Meloidogyne incognita*	EO; 1000 µg mL−1, Component: 100 & 200 µg mL−1 (48 h)	[54]
Cinnamomum zeylanicum	(e)-cinnamaldehyde, eugenol	J2 mortality, egg-hatching inhibition, reduction of galls and eggs	-	*Meloidogyne incognita*	EO; 49 µg/mL−1, Components: 529 &768 µg/mL−1 (48 h)	[54]
Tephrosia toxicaria	β-caryophyllene, germacrene D, a-humulene, bicyclogermacrene	Inhibition of J2 mortality and egg hatching	-	*Meloidogyne javanica; Meloidogyneenterolobii*	EO; 50, 100, 200, 400, 600, 800 µg mL−1 (48 h)	[55]
Cinnamomum cassia	(e)-cinnamaldehyde	J2 mortality, J2 paralysis, reduction of galls and eggs	Soybean	*Meloidogyne incognita*	EO; 62 µg/mL−1 Component: 208 µg/mL−1 (48 h)	[56]
Syzygium aromaticum	Eugenol**(Phenolics)**	J2 mortality	-	*Meloidogyne graminicola*	500 ppm (48 h)	[57]
Cymbopogon flexuosus	Citral	J2 mortality	-	*Meloidogyne graminicola*	Component: 500 ppm (48 h)	[57]
Pinus nigra	α-pinene, **(Terpene)**	J2 mortality	-	*Meloidogyne javanica*	EO; 1 µg mL−1 Compounds; 0.5 µg mL−1	[58]
Vetiveria zizanioides (L.)	**(Sesquiterpene)** 3,3,8,8-tetramethyltricyclo[5.1.0.0(2,4)]oct-5-ene-5-propanoic acid, 6-isopropenyl-4,8a-dimethyl-1,2,3,5,6,7,8,8aoctahydronaphthalen-2-ol	J2 mortality	-	*Meloidogyne graminicola*	EO; 0.95 mg/mL (72 h)	[59]
Citrus reticulata	limonene**(Monoterpene)**	J2 mortalityand egg-hatching inhibition	-	*Meloidogyne incognita*	1500 µg/mL−1 (96 h)	[60]
Hedychium coccineum	Spathulenol**(Sesquiterpene)**, Eucalyptol**(Monoterpene)**	J2 mortality and inhibition of egg-hatching	-	*Meloidogyne incognita*	EO; 0.25 µg/mL (24 h) 1 µg/mL (96 h)	[61]
-	Gallic acid **(Phenolics)**	-	Corrected mortality (Mc) induced by the phytochemicals4direct-contact bioassays	*Pratylenchus penetrans*	1.44 ± 0.41	[62]
-	Ferulic acid **(Phenolics)**	-	-	-	2.06 ± 0.34	[62]
-	Coumaric acid **(Phenolics)**	-	-	-	1.88 ± 0.37	[62]
-	Gentisic acid **(Phenolics)**	-	-	-	5.98 ± 1.30	[62]
-	Catechin **(Flavonoid)**	-	-	-	2.26 ± 0.61	[62]

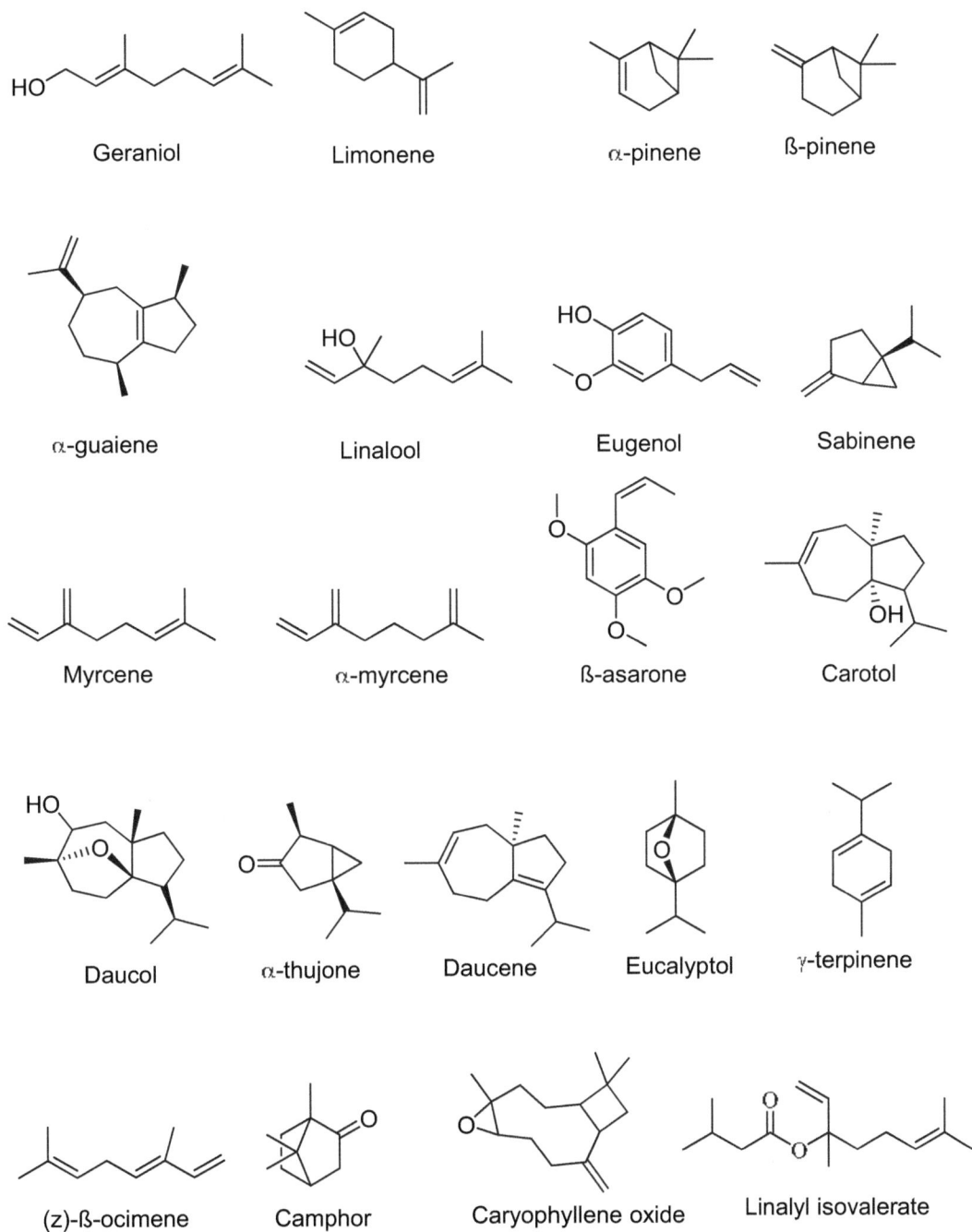

Fig. (1). Chemical structures of some of the phytochemical constituents with potent nematicidal activity.

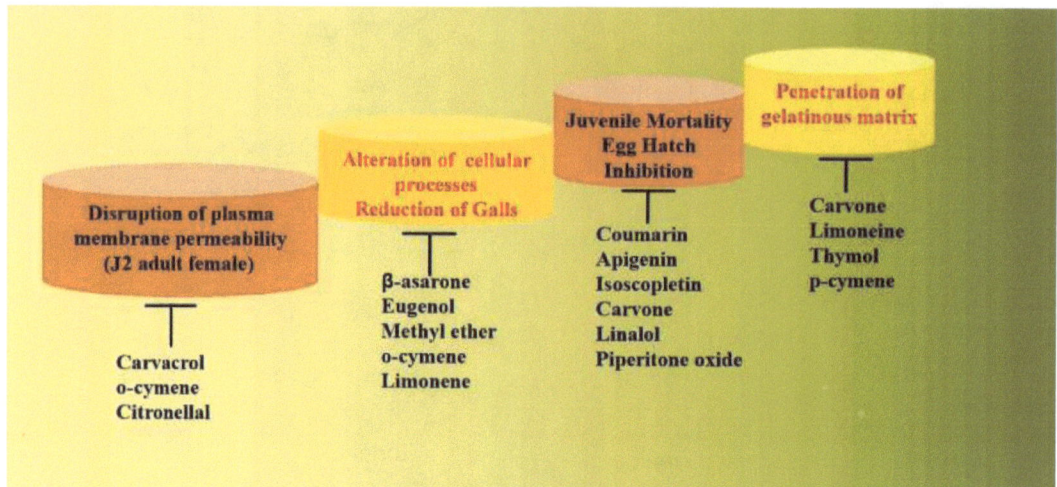

Fig. (2). Different physiological processes inhibited by various classes of plant secondary metabolites.

Terpenoids

Terpenoids a collective term used for compounds known as terpene (which is a 5-carbon-containing compound) and its derivatives, with 60,000 and identified as one of the largest secondary metabolite classes in plants [3]. Two units or more than two units of Activated dimethylallyl pyrophosphate or its isomer isopentenyl pyrophosphate are activated isoprene compounds that condense to form Terpenes [4]. On the basis of the total number of Carbon 5 building units involved, this condensation forms a Carbon 10 compound called monoterpene, a C15 compound (sesquiterpene), or a C20 compound (diterpene). Carbon 30 molecules known as triterpenes, *e.g.*, sterols, and Carbon 40 units called tetraterpenes *e.g.*, carotenoids can be formed by condensation of sesquiterpene and diterpene in a head-to-head manner [5]. Terpenoids can be formed by the substitution of terpenes by acetylation or hydroxylation. Since terpenoids are acted upon by enzymes that are numerous and highly diverse, hence terpenoids are the most diversified plant secondary metabolites class. Many members of the class of terpenoids have anti-pest and anti-pathogenic activity and the reason behind the increasing diversity of terpenoids seen throughout evolution could be attributed to an evolutionary arms race with these pathogens and pests [5].

In many varieties of *Capsicum annuum* (pepper), relative amounts of different terpenes in exudates of root were linked with the susceptibility to *M. incognita* [6]. Attractive or repellent effects of different terpenes released by the roots of *Capsicum annuum* on *M. incognita* J2s were confirmed by Olfactometer tests, which show that such terpenes play a part in plant pathogenic nematode by inhibiting or enhancing host-finding [6]. Terpenoid phytoalexins are found in different plant species, which are not yet evaluated for nematode resistance. To cite an example among cereals, three different classes of diterpenoid phytoalexins are produced in rice: phytocassanes, oryzalexins and momilactones; maize (*Zea mays*) produces sesquiterpenoid known as (zealexins) and diterpenoid (dolabralexins and kauralexins) [7, 8]. Such terpenoids are antifungal, antibacterial, and/or antipathogenic and anti-insect pests and these can be induced by treating with the resistance inducers which decrease sensitivity to nematodes hence it may be productive to investigate the role of phytoalexins in Plant Pathogenic Nematode resistance [5, 8 - 10]. Other than having direct effects against nematodes, terpenoids act as hormones in plants and thus play an indirect role in nematode-plant interactions. Brassinosteroids are the derivatives of triterpenoids, Abscisic acid is a derivative of tetraterpenoids, and gibberellic acid is a derivative of diterpenoids [11 - 13]. These hormones play numerous and often antagonistic roles in the resistance of plants to PPN [14 - 18].

Flavonoids

Flavonoids are one of the largest phenolic chemical compounds that belong to the category of plant secondary metabolites and have more than 10,000 identified members. Flavonoids play a role in the resistance of plants to diseases and pests apart from PPN. Flavonoids are one of the most extensively studied secondary metabolites w.r.t PPN resistance [19, 20]. The biosynthesis of flavonoids, initially is similar to the biosynthesis of stilbenoids, the difference lies in the enzyme used in the initial step, which is CHALCONE SYNTHASE and not STILBENE SYNTHASE. These enzymes have high sequence homology, and the enzyme STILBENE SYNTHASE is suggested to have eventually evolved from the enzyme CHALCONE SYNTHASE [21]. Three units of malonyl-CoA condense with *para*-coumaroyl-CoA units in the presence of the enzyme chalcone synthase to produce a compound known as chalcone skeleton, and then it further isomerizes to produce the respective flavonoid. Flavonoids can be further modified through *e.g.*, methylation, prenylation, glycosylation, and hydroxylation [22, 23]. Numerous commonly found flavonoids have been known to exhibit limited activity against nematodes, which has been deduced by *in vitro* experiments: kaempferol inhibits the hatching of eggs whereas myricetin and kaempferol quercetin, both are somewhat nematistatic (not nematicidal) and repellent to *M. incognita* juveniles [24]. Flavonoids show complex effects on nematode behaviour: Flavonoids either repel or attract *M. incognita* juveniles, which depends on their concentration and molecular structure [25]. Effects of these compounds on the interaction of plant and nematode have been widely studied in the Fabaceae family of plants whose members synthesize different isoflavonoids and pterocarpans (these are phytoalexins which are derived from compounds known as isoflavonoids *via* a combination of the isoflavonoid B ring at 4-one position) to counter the infection. Soriano *et al.* (2004) found that when *Avena sativa* (oat) plants were infected with *Pratylenchus neglectus* or *Heterodera avenae*, or if the plants were treated familiarly with methyl jasmonate which is a defense hormone, the quantity of 3 methanol-soluble compounds with UV-absorbance spectra resembling flavonoids increased 2-3 times in the roots and shoots [26]. A high nematicidal activity of the crude methanolic extract of methyl jasmonate-induced oat was observed against *H. avenae*. Out of the three flavonoid phytoalexins that can be induced, one significantly possessed nematicidal activity and the other two could be purified. Ultimately, three flavone-C-glycosides were discovered as inducible flavonoids: luteolin-C-hexoside-O-pentoside can not be purified, O-methyl-apigenin-C-deoxyhexoside-O-hexoside is nematicidal, and apigenin-C-hexoside-O-pentoside is not nematicidal. Apigenin isolated from *Artemisia incisa* showed the highest mortality and inhibition of *M. incognita* at the egg-hatching stage [27].

Tannins

Tannins comprise a diverse range of polyphenolic chemicals. They are further categorized into two subgroups: condensed tannins and hydrolyzable tannins. Condensed tannins are made up of oligomers of at least 2 or more flavan-3-ols, whereas esterification of Galloyl groups to a core of polyol is found in hydrolyzable tannins. Both subgroups of tannins show vast diversity in the extent of polymerization, along with other phenolic chemical compounds and monomer composition [28]. Tannins also play a role in resistance to herbivory by insects in plants and some studies have correlated nematode resistance and tannin accumulation [29].

Tannins are suggested to play a significant role in nematode resistance found in pinewood *B. xylophilus*. When *Bursaphelenchus xylophilus* was grown on the sap of phloem of eight different species of pine, the growth rate of *B. xylophilus* was observed to be negatively corresponding to the amount of the condensed tannins found in the phloem sap of studied species. Because of this, determining the proportionate share of flavonoids, phenolics, and condensed tannins-like compounds in the inhibitory impact on *B. xylophilus* is challenging [30].

CONCLUSION

Plant secondary metabolites exhibit great potential to protect commercial and agriculturally essential crops from the attack of plant parasitic nematodes. Phytoconstituents have shown great potential to kill the infective juveniles of PPNs to shield the plant from their action. Other researchers have documented similar studies that underscore the critical role of plant secondary metabolites in the management of infestations caused by plant-parasitic nematodes [63].

Additionally, plant-based formulations could be used as a potent and safe alternative to toxic synthetic pesticides. The need of the hour is to transition from using synthetically prepared pesticides towards organic and safer substitutes.

REFERENCES

[1] Kumar V, Khan MR, Walia RK. Crop loss estimations due to plant-parasitic nematodes in major crops in India. Natl Acad Sci Lett 2020; 43(5): 409-12.
[http://dx.doi.org/10.1007/s40009-020-00895-2]

[2] Hussein RA, El-Anssary AA. Plants secondary metabolites: the key drivers of the pharmacological actions of medicinal plants. Herbal medicine. 2019; 1(3): 11-30.
[http://dx.doi.org/10.5772/intechopen.76139]

[3] Pazouki L, Niinemets Ü. Multi-substrate terpene synthases: their occurrence and physiological significance. Front Plant Sci 2016; 7: 1019.
[http://dx.doi.org/10.3389/fpls.2016.01019] [PMID: 27462341]

[4] Cheng AX, Lou YG, Mao YB, Lu S, Wang LJ, Chen XY. Plant terpenoids: biosynthesis and ecological functions. J Integr Plant Biol 2007; 49(2): 179-86.
[http://dx.doi.org/10.1111/j.1744-7909.2007.00395.x]

[5] Pichersky E, Raguso RA. Why do plants produce so many terpenoid compounds? New Phytol 2018; 220(3): 692-702.
[http://dx.doi.org/10.1111/nph.14178] [PMID: 27604856]

[6] Kihika R, Murungi LK, Coyne D, *et al.* Parasitic nematode *Meloidogyne incognita* interactions with different *Capsicum annum* cultivars reveal the chemical constituents modulating root herbivory. Sci Rep 2017; 7(1): 2903.
[http://dx.doi.org/10.1038/s41598-017-02379-8] [PMID: 28588235]

[7] Yamane H. Biosynthesis of phytoalexins and regulatory mechanisms of it in rice. Biosci Biotechnol Biochem 2013; 77(6): 1141-8.
[http://dx.doi.org/10.1271/bbb.130109] [PMID: 23748776]

[8] Block AK, Vaughan MM, Schmelz EA, Christensen SA. Biosynthesis and function of terpenoid defense compounds in maize (*Zea mays*). Planta 2019; 249(1): 21-30.
[http://dx.doi.org/10.1007/s00425-018-2999-2] [PMID: 30187155]

[9] Lu X, Zhang J, Brown B, *et al.* Inferring roles in defense from metabolic allocation of rice diterpenoids. Plant Cell 2018; 30(5): 1119-31.
[http://dx.doi.org/10.1105/tpc.18.00205] [PMID: 29691314]

[10] Verbeek REM, Van Buyten E, Alam MZ, *et al.* Jasmonate-induced defense mechanisms in the belowground antagonistic interaction between *Pythium arrhenomanes* and *Meloidogyne graminicola* in rice. Front Plant Sci 2019; 10: 1515.
[http://dx.doi.org/10.3389/fpls.2019.01515] [PMID: 31824540]

[11] Choe S. Brassinosteroid biosynthesis and inactivation. Physiol Plant 2006; 126(4): 539-48.
[http://dx.doi.org/10.1111/j.1399-3054.2006.00681.x]

[12] Nambara E, Marion-Poll A. Abscisic acid biosynthesis and catabolism. Annu Rev Plant Biol 2005; 56(1): 165-85.
[http://dx.doi.org/10.1146/annurev.arplant.56.032604.144046] [PMID: 15862093]

[13] Hedden P, Thomas SG. Gibberellin biosynthesis and its regulation. Biochem J 2012; 444(1): 11-25.
[http://dx.doi.org/10.1042/BJ20120245] [PMID: 22533671]

[14] Nahar K, Kyndt T, Hause B, Höfte M, Gheysen G. Brassinosteroids suppress rice defense against root-knot nematodes through antagonism with the jasmonate pathway. Mol Plant Microbe Interact 2013; 26(1): 106-15.

[http://dx.doi.org/10.1094/MPMI-05-12-0108-FI] [PMID: 23194343]

[15] Kyndt T, Nahar K, Haeck A, Verbeek R, Demeestere K, Gheysen G. Interplay between carotenoids, abscisic acid and jasmonate guides the compatible *rice-Meloidogyne graminicola* interaction. Front Plant Sci 2017; 8: 951.
[http://dx.doi.org/10.3389/fpls.2017.00951] [PMID: 28642770]

[16] Bauters L, Hossain M, Nahar K, Gheysen G. Gibberellin reduces the susceptibility of rice, *Oryza sativa*, to the migratory nematode *Hirschmanniella oryzae*. Nematology 2018; 20(7): 703-9.
[http://dx.doi.org/10.1163/15685411-00003198]

[17] Song LX, Xu XC, Wang FN, *et al.* Brassinosteroids act as a positive regulator for resistance against root-knot nematode involving respiratory burst oxidase homolog-dependent activation of MAPKs in tomato. Plant Cell Environ 2018; 41(5): 1113-25.
[http://dx.doi.org/10.1111/pce.12952] [PMID: 28370079]

[18] Yimer HZ, Nahar K, Kyndt T, *et al.* Gibberellin antagonizes jasmonate-induced defense against *Meloidogyne graminicola* in rice. New Phytol 2018; 218(2): 646-60.
[http://dx.doi.org/10.1111/nph.15046] [PMID: 29464725]

[19] Mathesius U. Flavonoid functions in plants and their interactions with other organisms. Plants 2018; 7(2): 30.
[http://dx.doi.org/10.3390/plants7020030] [PMID: 29614017]

[20] Treutter D. Significance of flavonoids in plant resistance: a review. Environ Chem Lett 2006; 4(3): 147-57.
[http://dx.doi.org/10.1007/s10311-006-0068-8]

[21] Tropf S, Lanz T, Rensing SA, Schröder J, Schröder G. Evidence that stilbene synthases have developed from chalcone synthases several times in the course of evolution. J Mol Evol 1994; 38(6): 610-8.
[http://dx.doi.org/10.1007/BF00175881] [PMID: 8083886]

[22] Naoumkina MA, Zhao Q, Gallego-Giraldo L, Dai X, Zhao PX, Dixon RA. Genome-wide analysis of phenylpropanoid defence pathways. Mol Plant Pathol 2010; 11(6): 829-46.
[http://dx.doi.org/10.1111/j.1364-3703.2010.00648.x] [PMID: 21029326]

[23] Falcone Ferreyra ML, Rius SP, Casati P. Flavonoids: biosynthesis, biological functions, and biotechnological applications. Front Plant Sci 2012; 3: 222.
[http://dx.doi.org/10.3389/fpls.2012.00222] [PMID: 23060891]

[24] Wuyts N, Swennen R, De Waele D. Effects of plant phenylpropanoid pathway products and selected terpenoids and alkaloids on the behaviour of the plant-parasitic nematodes *Radopholus similis, Pratylenchus penetrans* and *Meloidogyne incognita*. Nematology 2006; 8(1): 89-101.
[http://dx.doi.org/10.1163/156854106776179953]

[25] Kirwa HK, Murungi LK, Beck JJ, Torto B. Elicitation of differential responses in the root-knot nematode *Meloidogyne incognita* to tomato root exudate cytokinin, flavonoids, and alkaloids. J Agric Food Chem 2018; 66(43): 11291-300.
[http://dx.doi.org/10.1021/acs.jafc.8b05101] [PMID: 30346752]

[26] Soriano IR, Asenstorfer RE, Schmidt O, Riley IT. Inducible flavone in oats (*Avena sativa*) is a novel defense against plant-parasitic nematodes. Phytopathology 2004; 94(11): 1207-14.
[http://dx.doi.org/10.1094/PHYTO.2004.94.11.1207] [PMID: 18944456]

[27] Khan R, Naz I, Hussain S, *et al.* Phytochemical management of root knot nematode (*Meloidogyne incognita*) kofoid and white chitwood by *Artemisia* spp. in tomato (*Lycopersicon esculentum* L.). Braz J Biol 2020; 80(4): 829-38.
[http://dx.doi.org/10.1590/1519-6984.222040] [PMID: 31800766]

[28] Barbehenn RV, Peter Constabel C. Tannins in plant–herbivore interactions. Phytochemistry 2011; 72(13): 1551-65.

[http://dx.doi.org/10.1016/j.phytochem.2011.01.040] [PMID: 21354580]

[29]   Salminen JP, Karonen M. Chemical ecology of tannins and other phenolics: we need a change in approach. Funct Ecol 2011; 25(2): 325-38.
[http://dx.doi.org/10.1111/j.1365-2435.2010.01826.x]

[30]   Pimentel CS, Firmino PN, Calvão T, Ayres MP, Miranda I, Pereira H. Pinewood nematode population growth in relation to pine phloem chemical composition. Plant Pathol 2017; 66(5): 856-64.
[http://dx.doi.org/10.1111/ppa.12638]

[31]   Elsharkawy MM, Al-Askar AA, Behiry SI, *et al.* Resistance induction and nematicidal activity of certain monoterpenes against tomato root-knot caused by *Meloidogyne incognita.* Front Plant Sci 2022; 13: 982414.
[http://dx.doi.org/10.3389/fpls.2022.982414] [PMID: 36204064]

[32]   Bu MM, Yu SQ, Dong CZ. [Chemical constituents from fruits of *Aristolochia mollissima* and their nematicidal activity against root-knot nematode]. Zhongguo Zhongyao Zazhi 2018; 43(16): 3307-14.
[PMID: 30200734]

[33]   Kiran Z, Khan HN, Rasheed S, *et al.* Isolation of secondary metabolites from *Syzygium aromaticum* (L.) Merr. & L.M.Perry. (cloves), and evaluation of their biological activities. Nat Prod Res 2023; 37(12): 2018-23.
[http://dx.doi.org/10.1080/14786419.2022.2112956] [PMID: 35997246]

[34]   D'Addabbo T, Laquale S, Argentieri MP, Bellardi MG, Avato P. Nematicidal activity of essential oil from lavandin (*Lavandula× intermedia Emeric* ex Loisel.) as related to chemical profile. Molecules 2021; 26(21): 6448.
[http://dx.doi.org/10.3390/molecules26216448] [PMID: 34770856]

[35]   Laquale S, Avato P, Argentieri MP, Bellardi MG, D'Addabbo T. Nematotoxic activity of essential oils from *Monarda* species. J Pest Sci 2018; 91(3): 1115-25.
[http://dx.doi.org/10.1007/s10340-018-0957-1]

[36]   Gowda A, Shakil NA, Rana VS, Singh AK, Bhatt KC, Devaraja KP. Chemical composition and nematicidal activity of essential oil and piperitenone oxide of *Mentha longifolia* L. against *Meloidogyne incognita.* Allelopathy J 2023; 58(2): 165-82.
[http://dx.doi.org/10.26651/allelo.j/2023-58-2-1427]

[37]   Hammad EA, Hasanin MMH. Antagonistic Effect of Nanoemulsions of Some Essential Oils against Fusarium oxysporum and Root-Knot Nematode *Meloidogyne javanica* on Coleus Plants. Pak J Nematol 2022; 40(1)
[http://dx.doi.org/10.17582/journal.pjn/2022/40.1.35.48]

[38]   Hammad EA, El-Sagheer AM. Comparative efficacy of essential oil nanoemulsions and bioproducts as alternative strategies against root-knot nematode, and its impact on the growth and yield of *Capsicum annuum* L. J Saudi Soc Agric Sci 2023; 22(1): 47-53.
[http://dx.doi.org/10.1016/j.jssas.2022.06.002]

[39]   Eloh K, Kpegba K, Sasanelli N, Koumaglo HK, Caboni P. Nematicidal activity of some essential plant oils from tropical West Africa. Int J Pest Manage 2020; 66(2): 131-41.
[http://dx.doi.org/10.1080/09670874.2019.1576950]

[40]   Felek AF, Ozcan MM, Akyazi F. Effects of essential oils distilled from some medicinal and aromatic plants against root knot nematode (*Meloidogyne hapla*). J Appl Sci Environ Manag 2019; 23(8): 1425-30.
[http://dx.doi.org/10.4314/jasem.v23i8.3]

[41]   Keerthiraj M, Mandal A, Dutta TK, *et al.* Nematicidal and molecular docking investigation of essential oils from *Pogostemon cablin* ecotypes against *Meloidogyne incognita.* Chem Biodivers 2021; 18(9): e2100320.
[http://dx.doi.org/10.1002/cbdv.202100320] [PMID: 34245651]

[42]   Kundu A, Dutta A, Mandal A, *et al.* A comprehensive *in vitro* and *in silico* analysis of nematicidal action of essential oils. Front Plant Sci 2021; 11: 614143.
[http://dx.doi.org/10.3389/fpls.2020.614143] [PMID: 33488658]

[43]   Ntalli N, Bratidou Parlapani A, Tzani K, *et al. Thymus citriodorus* (Schreb) botanical products as ecofriendly nematicides with bio-fertilizing properties. Plants 2020; 9(2): 202.
[http://dx.doi.org/10.3390/plants9020202] [PMID: 32041220]

[44]   Ardakani AS, Hosseininejad SA. Identification of chemical components from essential oils and aqueous extracts of some medicinal plants and their nematicidal effects on *Meloidogyne incognita*. J Basic Appl Zool 2022; 83(1): 14.
[http://dx.doi.org/10.1186/s41936-022-00279-6]

[45]   Kabdal T, Himani , Kumar R, *et al.* Seasonal variation in the essential oil composition and biological activities of *Thymus linearis* Benth. Collected from the Kumaun region of Uttarakhand, India. Biochem Syst Ecol 2022; 103: 104449.
[http://dx.doi.org/10.1016/j.bse.2022.104449]

[46]   Ferreira Barros A, Paulo Campos V, Lopes de Paula L, *et al.* The role of *Cinnamomum zeylanicum* essential oil, (*E*)-cinnamaldehyde and (*E*)-cinnamaldehyde oxime in the control of *Meloidogyne incognita*. J Phytopathol 2021; 169(4): 229-38.
[http://dx.doi.org/10.1111/jph.12979]

[47]   Galisteo A, González-Coloma A, Castillo P, Andrés MF. Valorization of the hydrolate byproduct from the industrial extraction of purple *Alium sativum* essential oil as a source of nematicidal products. Life (Basel) 2022; 12(6): 905.
[http://dx.doi.org/10.3390/life12060905] [PMID: 35743936]

[48]   Pardavella I, Daferera D, Tselios T, Skiada P, Giannakou I. The use of essential oil and hydrosol extracted from *Cuminum cyminum* seeds for the control of *Meloidogyne incognita* and *Meloidogyne javanica*. Plants 2021; 10(1): 46.
[http://dx.doi.org/10.3390/plants10010046] [PMID: 33379232]

[49]   Kaur A, Chahal KK, Kataria D, Urvashi KA. Assessment of carrot seed essential oil and its chemical constituents against *Meloidogyne incognita*. J Pharmacogn Phytochem. 2018; 7(1): 896-903.

[50]   Basaid K, Chebli B, Bouharroud R, *et al.* Biocontrol potential of essential oil from Moroccan *Ridolfia segetum* (L.) Moris. J Plant Dis Prot 2021; 128(5): 1157-66.
[http://dx.doi.org/10.1007/s41348-021-00489-0]

[51]   Kalaiselvi D, Mohankumar A, Shanmugam G, Thiruppathi G, Nivitha S, Sundararaj P. Altitude-related changes in the phytochemical profile of essential oils extracted from *Artemisia nilagirica* and their nematicidal activity against *Meloidogyne incognita*. Ind Crops Prod 2019; 139: 111472.
[http://dx.doi.org/10.1016/j.indcrop.2019.111472]

[52]   Ismail M, Kowsar A, Javed S, *et al.* The antibacterial, insecticidal and nematocidal activities and toxicity studies of *Tanacetum falconeri* Hook. f. Turk J Pharm Sci 2021; 18(6): 744-51.
[http://dx.doi.org/10.4274/tjps.galenos.2021.63372] [PMID: 34979736]

[53]   Dutta A, Mandal A, Kundu A, *et al.* Deciphering the behavioral response of *Meloidogyne incognita* and Fusarium oxysporum toward mustard essential oil. Front Plant Sci 2021; 12: 714730.
[http://dx.doi.org/10.3389/fpls.2021.714730] [PMID: 34512695]

[54]   Barros AF, Campos VP, de Oliveira DF, *et al.* Activities of essential oils from three Brazilian plants and benzaldehyde analogues against *Meloidogyne incognita*. Nematology 2019; 21(10): 1081-9.
[http://dx.doi.org/10.1163/15685411-00003276]

[55]   Moreira FJC, Araújo BA, Lopes FGN, *et al.* Assessment of the *Tephrosia toxicaria* essential oil on hatching and mortality of eggs and second-stage juvenile (J2) root-knot nematode (*Meloidogyne enterolobii* and *M. javanica*). Aust J Crop Sci 2018; 12(12): 1829-36.
[http://dx.doi.org/10.21475/ajcs.18.12.12.p1102]

[56]   Jardim IN, Oliveira DF, Silva GH, Campos VP, de Souza PE. (E)-cinnamaldehyde from the essential oil of *Cinnamomum cassia* controls *Meloidogyne incognita* in soybean plants. J Pest Sci 2018; 91(1): 479-87.
[http://dx.doi.org/10.1007/s10340-017-0850-3]

[57]   Ajith M, Pankaj , Shakil NA, Kaushik P, Rana VS. Chemical composition and nematicidal activity of essential oils and their major compounds against *Meloidogyne graminicola* (Rice Root-Knot Nematode). J Essent Oil Res 2020; 32(6): 526-35.
[http://dx.doi.org/10.1080/10412905.2020.1804469]

[58]   Mamoci E, Andrés MF, Olmeda S, González-Coloma A. Chemical composition and activity of essential oils of Albanian coniferous plants on plant pests. Chemistry Proceedings 2022; 10(1): 15.

[59]   Jindapunnapat K, Reetz ND, MacDonald MH, *et al.* Activity of vetiver extracts and essential oil against *Meloidogyne incognita*. J Nematol 2018; 50(2): 147-62.
[http://dx.doi.org/10.21307/jofnem-2018-008] [PMID: 30451435]

[60]   Goyal L, Kaushal S, Dhillon NK, Heena . Nematicidal potential of *Citrus reticulata* peel essential oil, isolated major compound and its derivatives against *Meloidogyne incognita*. Arch Phytopathol Pflanzenschutz 2021; 54(9-10): 449-67.
[http://dx.doi.org/10.1080/03235408.2021.1890369]

[61]   Arya S, Kumar R, Prakash O, *et al.* Chemical composition and biological activities of *Hedychium coccineum* Buch.-Ham. ex Sm. essential oils from Kumaun hills of Uttarakhand. Molecules 2022; 27(15): 4833.
[http://dx.doi.org/10.3390/molecules27154833] [PMID: 35956784]

[62]   Barbosa P, Faria JMS, Cavaco T, Figueiredo AC, Mota M, Vicente CSL. Nematicidal activity of phytochemicals against the root-lesion nematode *Pratylenchus penetrans*. Plants 2024; 13(5): 726.
[http://dx.doi.org/10.3390/plants13050726] [PMID: 38475572]

[63]   Sarri K, Mourouzidou S, Ntalli N, Monokrousos N. Recent advances and developments in the nematicidal activity of essential oils and their components against root-knot nematodes. Agronomy (Basel) 2024; 14(1): 213.
[http://dx.doi.org/10.3390/agronomy14010213]

CHAPTER 5

Mode of Action of Phytochemicals During Physiological and Biochemical Interactions with Nematodes

Istkhar Rao[1,*], Kajol Yadav[1] and Aashaq Hussain Bhat[2]

[1] *Department of Biosciences and Biotechnology, Banasthali Vidyapith, Banasthali 304022, Rajasthan, India*

[2] *Department of Bioscience, University Centre for Research and Development, Chandigarh University, Gharuan, Mohali 140413, Punjab, India*

Abstract: The proliferation of plant-parasitic nematodes as formidable agricultural pests poses a significant global threat to crop productivity. Despite their diminutive size, these organisms inflict substantial economic losses, with global damage surpassing that caused by insect pests. The cryptic nature of nematode infections renders them particularly insidious, often leading to underestimation and inadequate management. Beyond their intrinsic harmful effects, plant-parasitic nematodes exacerbate crop damage by forming synergistic disease complexes with other pathogenic microorganisms. Nematodes utilize diverse strategies to breach plant host tissues, with a particular emphasis on the root-knot and cyst-forming nematodes—two prominent groups that inflict severe agricultural damage. The evolution of plant defense mechanisms is an intrinsic biological response by which plants counteract nematode parasitism. Plants deploy receptor molecules against nematode effectors, facilitating resistance by either preventing nematode penetration or by producing nematicidal proteins that mitigate nematode pathogenicity. The activation of plant defense-related genes and the synthesis of defensive hormones are pivotal in enhancing plant resilience against nematode invasion. However, under certain conditions, these defensive strategies may inadvertently augment nematode parasitism. Common symptoms indicative of nematode infestation include tissue necrosis, gall formation, cyst development, and stunted plant growth. This chapter delves into the current understanding of plant-nematode interactions, emphasizing the molecular and physiological mechanisms underpinning plant immune responses to nematode invasion.

Keywords: Effector molecules, Nematode feeding sites, Plant hormones, Plant parasitic nematodes, Secondary metabolites.

* **Corresponding author Istkhar Rao:** Department of Biosciences and Biotechnology, Banasthali Vidyapith, Banasthali 304022, Rajasthan, India; E-mails: istkharrao@gmail.com, istkhar@banasthali.in

Shivam Jasrotia & Ajay Kumar (Eds.)

INTRODUCTION

Nearly all life forms on Earth, including humans and other animals, are heavily reliant on plants as their primary source of energy. The escalating human population has necessitated the cultivation of plants to meet its growing needs. Consequently, a substantial portion of Earth's terrestrial landscape has been repurposed by humans to cultivate crops essential for human sustenance, leading to significant advancements in food production. However, since Earth is inhabited by a multitude of organisms, not just humans, the reliability of continuous food production remains uncertain. This is partly due to the presence of pests—defined here as any organism, including pathogens—that can detrimentally impact plant growth, defense mechanisms, and nutritional quality [1, 2]. Annually, pests are responsible for over 40% of global crop losses (IPPC Secretariat, 2021). While the term 'pests' encompasses a broad range of organisms, these entities primarily compete with humans for food and habitat. Both invertebrate and vertebrate pests inflict considerable damage to crops. Although pesticides are commonly employed for pest management, their application often entails substantial costs, including biodiversity loss and ecological harm, which may outweigh the immediate benefits of crop protection [3]. Among the myriad pests, insects, and nematodes are recognized as two of the most prominent groups responsible for significant crop yield reductions worldwide.

NEMATODES AS PLANT PARASITES

Nematodes exhibit extraordinary adaptability, thriving in diverse environments such as deserts, marshes, oceans, tropical regions, and even Antarctica, as they have been reported from every continent. These organisms encompass a variety of life forms, including free-living parasites, fungivores, bacterivores, and even nematophagous (nematode-eating) species. Despite their diminutive size and the challenge of species identification, nematodes hold a pivotal role in ecosystems. Estimates suggest that the global nematode species count could exceed 500,000, although only about 29,000 species have been formally described to date [4]. Their diversity and abundance are second only to insects among multicellular organisms. While many nematodes are parasitic to plants and animals, a significant number are free-living and feed on bacteria, fungi, protozoans, and other nematodes. Currently, there are approximately 4,100 recognized plant parasitic nematode (PPN) species, primarily targeting plant roots, though some also infest aerial plant parts. These PPNs account for roughly 15% of all known nematode species [5, 6]. Although nematode diversity is vast, only a few species cause substantial agricultural damage. Jones *et al.* [7] identified the 11 most critical nematode genera affecting crops, including *Meloidogyne, Heterodera, Globodera, Pratylenchus, Radopholus, Ditylenchus, Bursaphelenchus,*

Rotylenchulus, *Xiphinema*, *Nacobbus*, and *Aphelenchoides*. Collectively, these PPN species are responsible for annual agricultural losses exceeding $358 billion worldwide [8].

To manage these pests, large quantities of nematicides are used. The global nematicide market was estimated at $1.58 billion in 2022, with projections suggesting it could grow to $4.28 billion by 2032. The Asia Pacific region is expected to experience the fastest market growth during this period [https://www.sphericalinsights.com]. Effective nematode management is highly challenging due to the difficulty in diagnosing above-ground symptoms of nematode infestation. Therefore, an integrated management approach that combines chemical treatments, cultural practices, biological control, and resistant plant varieties is essential [5]. Numerous nematicides, such as aldicarb, carbofuran, fenamiphos, oxamyl, and methyl bromide, have been employed to control nematode populations by disrupting their nervous systems, causing paralysis and death. Despite the challenges associated with commercializing bionematicides, various biopesticides, particularly those utilizing bacterial and fungal antagonists, have shown promise. Plant nematologists have invested considerable effort into understanding the mechanisms of plant defence against PPNs, including the synthesis of metabolites with anti-nematode properties, known as anti-nematode phytochemicals (ANPs).

Plant parasitic nematodes possess unique abilities to detect and respond to chemical signals from their hosts, enabling them to locate and infect plant tissues. Over evolutionary time, plants and nematodes have co-evolved, developing mechanisms to either parasitize or defend against each other. Understanding the complex molecular signalling and interactions during the early stages of host-parasite relationships is crucial for identifying vulnerable points in the nematode lifecycle and disrupting nematode-host recognition. Nematode infestations can significantly reduce crop yields by invading plant roots, altering root architecture, and diminishing the plant's ability to absorb nutrients and water.

FEEDING STRATEGIES OF PLANT PARASITIC NEMATODES

Diagnosing nematode infections based solely on above-ground symptoms can be challenging, as these symptoms often mimic those of nutrient deficiencies. Plant parasitic nematodes primarily feed on plant roots, but they can also parasitize stems, leaves, flowers, and seeds. Despite their diverse feeding strategies, all nematodes utilize a specialized structure called a stylet, located in their stoma, to pierce plant tissues. The stylet's morphology is critical for taxonomic identification and provides insight into the nematode's feeding habits. Fewer nematodes feed on above-ground plant parts such as leaves, stems, and bulbs.

Notable nematode genera in this category include *Bursaphelenchus* (pine wood nematodes), *Ditylenchus* (stem and bulb nematodes), *Anguina* (seed gall nematodes), and *Aphelenchoides* (foliar nematodes). A larger proportion of nematodes parasitize root tissues and are classified into ectoparasites, semi-endoparasites, migratory endoparasites, and sedentary endoparasites. Examples of genera within these groups include *Belonolaimus*, *Xiphinema*, *Trichodorus*, *Rotylenchulus*, *Tylenchulus*, *Pratylenchus*, *Radopholus*, *Meloidogyne*, *Heterodera*, and *Naccobus*. Table **1** below provides a summary of the key characteristics of these nematode genera about their interactions with plants.

Table 1. Important Feeding strategies followed by plant parasitic nematodes and their feeding behaviour in different plant crops.

Feeding	Genera	Common Name	Order	Feeding	Affected Crops	References
Ectoparasite	*Belonolaimus*	Sting Nematode	Tylenchida	All stages are parasitic and feed from outside the root.	Turfgrasses, forage grasses, strawberries, potatoes, sugarcane, cantaloupe, peanuts, cabbage, cotton	[8, 9]
	Xiphenema	Dagger nematodes	Dorylaimida	All stages are parasitic and feed from outside the root.	Grape, strawberry, hops, vegetables	[10]
	Paratrichodorus	Stubby root nematode	Triplonchida	All stages are parasitic and feed from outside the root.	Potato, tobacco, narcissus, tulip, aster, gladiolus, hyacinth, and sugar beet	[11]
	Trichodorus	Stubby-root Nematodes	Triplonchida	All stages are parasitic and feed from outside the root.	Corn, soybean, sorghum, peanut, turf grasses, potato, and other vegetables	[12]

(Table 1) cont.....

Feeding	Genera	Common Name	Order	Feeding	Affected Crops	References
Semi-Endoparasites	*Rotylenchulus*	Reniform Nematodes	Tylenchida	Partially inside the root.	Cotton, pineapple, and many vegetable crops including tomato, okra, squash, and lettuce	[13]
	Sphaeronema	-	Tylenchida	All stages are parasitic and feed from partially inside the root.	Red spruce and Fraser fir	[14]
	Helicotylenchus	Spiral nematode	Tylenchida	All stages are parasitic and feed from inside the root.	Banana plantains, soybeans, cotton, and corn	[15, 16]
	Tylenchulus	Citrus nematodes	Tylenchida	Ectoparasite as a juvenile and young adult female, sedentary semi-endoparasite as a mature female.	Citrus, trifoliate orange, grapevines, persimmon, lilac, and olive	[17]
Migratory Endoparasites	*Pratylenchus*	Lesion Nematodes	Tylenchida	All stages are parasitically fed and move in and out of roots.	Soybeans, potatoes, corn, bananas, date palm, oil palm, peanuts and wheat	[18]
	Hoplolaimus	Lance Nematodes	Tylenchida	J2 to adult and feed from the root and shoot.	Sugarcane, soybean, cabbages, cauliflowers	[19]
	Radopholus	Burrowing Nematodes	Tylenchida	All stages are parasitically fed and move in and out of roots.	Banana	[20, 21]
	Hirschmanniella	Root-lesion nematodes	Tylenchida	All juvenile and adult stages migrate through the root cortex and feed on cells.	Rice and date palm	[22]

(Table 1) cont.....

Feeding	Genera	Common Name	Order	Feeding	Affected Crops	References
Sedentary Endoparasites	*Meloidogyne*	Root-knot Nematodes	Tylenchida	Juveniles penetrate the root and molt into females, which become sedentary inside the roots. Males are rare.	Tomato, okra, eggplant, carrot, guava, sweet potato, rice and legume crops	[23]
	Heterodera	Cyst Nematodes	Tylenchida	Male and female Juveniles penetrate the roots. Females become sedentary inside the roots.	Wheat, barley, pigeon pea, cowpea, moonbeam, soybean, black gram, sesame, oat, ray	[24]
	Nacobbus	False Root-Knot Nematodes	Tylenchida	Juvenile stages are migratory endoparasites, Females are sedentary.	Potato, tomato, sugar beet, carrot, lettuce, spinach, pepper and cucumber	[25, 26]
Stems	*Bursaphelenchus*	Pine Wood Nematodes	Aphelenchida	Migratory endoparasite enter the pine through its vector beetle and enter inside the plant as a J4 stage.	Pine trees	[27]
	Ditylenchus	Stem and Bulb Nematodes	Tylenchida	Migratory ectoparasites and endoparasites of stems, bulbs, and tubers. J4 enters the plant through young tissues and seedlings.	Onion, garlic, potato	[28]

Feeding	Genera	Common Name	Order	Feeding	Affected Crops	References
Seed	*Anguina*	Seed Gall Nematodes	Tylenchida	The juvenile state (J2) enters through roots, and with the help of a film of water reaches the stem growing tip and probably feeds ectoparasitically.	Wheat, oats, barley, rye	[30]
Foliar part	*Aphelenchoides*	Foliar Nematodes	Aphelenchida	Ectoparasite. All stages are parasitic.	Rice strawberry, tuberose	[31]

PHYSIOLOGY OF PLANT NEMATODE INTERACTION

Plant-parasitic nematodes inflict damage on plants through various mechanisms, primarily by triggering basal immune responses upon initial contact with host plant roots. Root infection is the most prevalent form of nematode invasion, typically involving the penetration and entry of second-stage juveniles (J2) into the host. These juveniles secrete a diverse array of effector molecules to facilitate entry. Traditionally, the term "effector molecule" was confined to proteins secreted by pathogens, but it now encompasses peptides, small RNAs, and secondary metabolites as well [32 - 34]. Upon secretion and subsequent interaction with host tissues, these effectors initiate the degradation of host cell walls, thereby creating pathways for nematode ingress. Secretions from subventral and dorsal glands—modifications of pharyngeal gland cells—are thought to be critical in both the initial parasitism during early developmental stages and in sustaining the sedentary phase later in the nematode's lifecycle [35, 36].

Lesion nematodes, including those from the genera *Pratylenchus*, *Radopholus*, *Hirschmanniella*, and *Bursaphelenchus*, cause extensive necrotic damage as they migrate through host plant tissues [37]. Root-knot nematodes (*Meloidogyne* spp.) and cyst nematodes (*Globodera* spp. and *Heterodera* spp.), classified as sedentary endoparasites due to their establishment and immobility within host tissues, significantly impair plant growth by extracting nutrients from the roots [7]. In cyst nematodes, females become sessile upon reaching maturity, while males remain motile in the soil, resembling filaments. Under stress conditions—such as suboptimal plant growth or nutrient limitations—some J2 juveniles of *Meloidogyne* spp. differentiate into males rather than females, mitigating resource competition as males exit the root without feeding. A primary distinction between these nematode groups is the site of root penetration; cyst nematodes penetrate at poorly defined sites, whereas root-knot nematodes typically invade just behind the root tip in the elongation zone. Additionally, their migratory paths differ; cyst nematodes traverse intercellular spaces, while root-knot nematodes advance towards the vascular cylinder.

Upon successful entry, nematodes establish feeding sites by manipulating host plant physiology, inducing species-specific changes. For instance, in cyst-forming nematodes like *Heterodera* and *Globodera*, the introduction of effector molecules results in the formation of a syncytium at the feeding site. This condition induces significant cellular changes, including cell cycle arrest at the G_2 phase, subcellular component proliferation, and fusion of adjacent cells due to the digestion of the middle lamella and enlargement of plasmodesmata. This process culminates in a multinucleated nutrient sink, exploited by nematodes for their growth and development [38, 39]. Conversely, *Meloidogyne* spp. (root-knot nematodes) do not form syncytia; instead, they stimulate the formation of giant cells by selecting multiple cells for feeding, which continue to undergo mitotic division without cytokinesis, resulting in large, multinucleated cells. Surrounding these giant cells, the neighbouring cells proliferate to form galls, providing a habitat for the nematodes [40].

Successful parasitism initiates alterations in the nematode's developmental stages, with J2 larvae molting into J3, and subsequently into J4. Further progression leads to sexual maturation, influenced by environmental conditions. Under adverse conditions, such as those presented by resistant plant varieties, there is an increased tendency for juveniles to develop into males [41, 42]. An overview of the life cycle of plant-parasitic nematodes and the specific chemicals involved in their interactions with host plants is depicted in Fig. (**1**).

GENETIC, MOLECULAR, AND BIOCHEMICAL INTERACTIONS BETWEEN PLANTS AND NEMATODES

The majority of current nematode pest management strategies are considered obsolete, with many synthetic nematicides having been restricted or outright banned due to their adverse effects on both the environment and human health. Therefore, the development of novel, less harmful strategies for controlling plant-parasitic nematodes (PPNs) that align with sustainable agricultural practices is of paramount importance. In this context, understanding the genetic interactions between plants and nematodes is critical for identifying innovative approaches to nematode management that can enhance crop productivity, thereby addressing issues of food security and mitigating economic losses. There is no doubt that contemporary approaches to mitigating nematode infestations, such as deploying resistant cultivars, implementing crop rotation schemes, and utilizing biological control methods, represent some of the most promising strategies [43].

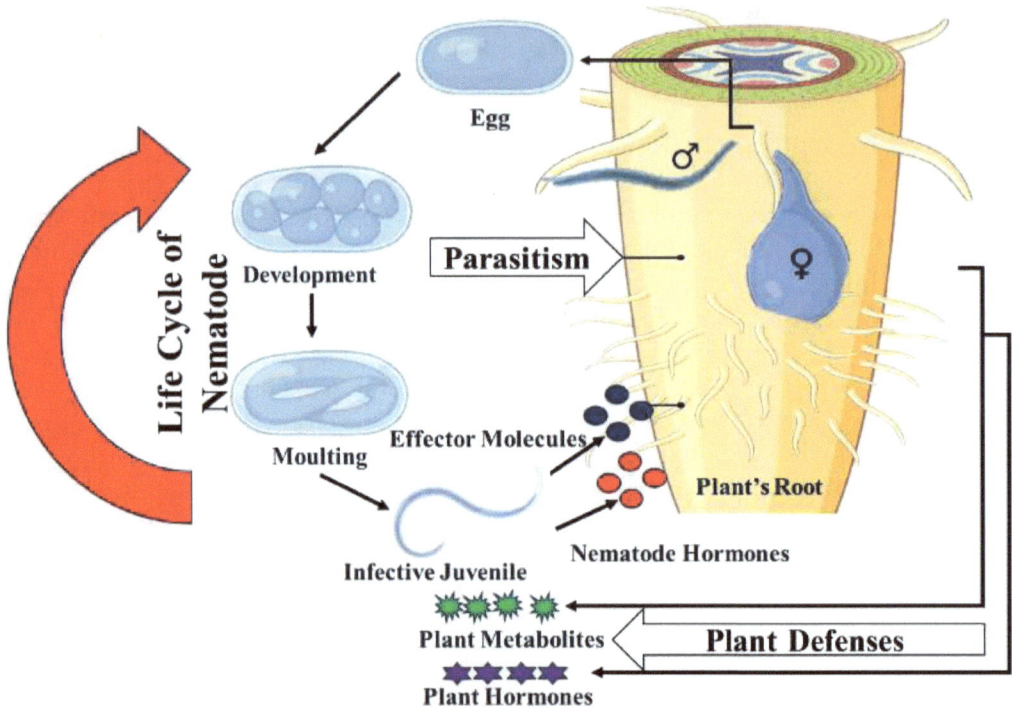

Fig. (1). Depiction of selected molecules involved in plant nematode interaction and life cycle of an endoparasitic nematode.

Plant Immune Responses to Nematode Parasitism

Upon contact with nematodes, plants activate their immune defence mechanisms. Two principal types of plant immune responses have been identified: PAMP/MAMP (pathogen-associated molecular patterns or microbial-associated molecular patterns)-triggered immunity (PTI/MTI) and effector-triggered immunity (ETI). In the context of nematode interactions, these PAMPs are referred to as nematode-associated molecular patterns (NAMPs) [44]. Pattern recognition receptors (PRRs) on the plant cell surface initiate MTI, whereas nucleotide-binding domain leucine-rich repeat proteins (NB-LRRs) within the host cell recognize specific effector molecules secreted by the invading organism, thereby activating ETI [45].

Phytochemicals and Plant Metabolites

Analogous to animal metabolites, plant metabolites can be broadly categorized into primary and secondary classes. Primary metabolites are directly involved in essential physiological processes such as growth, development, and cellular differentiation. In contrast, secondary metabolites, while not indispensable for basic plant development, are synthesized during metabolic processes and play vital roles in plant health, including pollinator attraction, abiotic stress tolerance, and defence against pests and pathogens [46]. However, this classification is sometimes ambiguous, as certain metabolites, like plant hormones and lignin monomers, exhibit dual roles, functioning both as primary and secondary metabolites [47]. For example, shikimic acid, integral to the biosynthesis of aromatic amino acids and vitamins, and squalene, a precursor in steroid biosynthesis, have been reclassified from secondary to primary metabolites due to their fundamental biological functions [47].

It is estimated that approximately 200,000 distinct types of secondary metabolites are synthesized through extensive secondary metabolic pathways in plants [48]. These bioactive compounds, often termed phytochemicals, are crucial for plant defence mechanisms. Notable phytochemicals include dietary fibres, various polysaccharides, carotenoids, polyphenols, isoprenoids, phytosterols, and saponins, which possess antiviral, antibacterial, antidiarrheal, anthelmintic, antiallergic, and potent antioxidant properties [49, 50]. These phytochemicals are typically sourced from diverse plant-based foods such as whole grains, fruits, vegetables, nuts, and herbs.

Secondary metabolites are further categorized into phytoanticipins and phytoalexins. Phytoanticipins are pre-formed antimicrobial compounds produced during normal plant growth that inhibit pathogen development. These compounds may remain in vacuoles as inactive substances, such as quinones, catechol, and protocatechuic acid in onions, or be released into the surrounding environment, including the rhizosphere [51]. These defence molecules are constitutively present, independent of pathogenic invasion [52]. Conversely, phytoalexins are low molecular weight antimicrobial and antioxidative compounds synthesized *de novo* in response to biotic and abiotic stressors, rapidly accumulating at the sites of pathogen attack [52]. The distinction between phytoanticipins and phytoalexins is not always clear-cut, as many phytochemicals may function as both, depending on specific plant species and environmental conditions. For instance, resveratrol, commonly recognized as a phytoalexin, is classified as a phytoanticipin in certain plant species [53]. Similarly, S-methyl-cysteine sulfoxide, initially identified as a phytoalexin, is now regarded as a phytoanticipin due to its broader antimicrobial activity [54, 55]. It is imperative to understand that the differentiation between phytoalexins and phytoanticipins is based on their mode of synthesis rather than their chemical composition.

Various plants produce distinct metabolites that function as nematode antagonists. For instance, the phenolic compound chlorogenic acid, although present in various plants, exhibits limited nematicidal activity against *Nacobbus aberrans* and *Meloidogyne incognita* [56 - 58]. Conversely, the phenylphenalenone and anigorufone, which accumulates around the roots of resistant banana cultivars infected by *Radopholus similis*, demonstrates significant nematicidal, antifungal (against *Fusarium oxysporum*), and antiprotozoal (against *Leishmania*, the causative agent of leishmaniasis) activities [59]. Flavonoids, a major group of secondary metabolites, include flavonols such as kaempferol, quercetin, and myricetin, which exhibit nemastatic properties by preventing nematodes from approaching host plant roots or delaying nematode egg hatching [60]. Other flavonoids, such as patuletin, patulitrin, quercetin, and rutin, are nematicidal against *Heteroderazeae* juveniles [61], while medicarpin (an isoflavonoid) inhibits the motility of *Pratylenchus penetrans* [62], and glyceollins (pterocarpans) deter the motility of *M. incognita* [63]. According to Hamaguchi *et al.* [64], α-terthienyl derived from marigold roots exhibits nematicidal activity by penetrating the nematode hypodermis. Similarly, asparagus-derived compounds suppress the hatching of *Globodera rostochiensis* and *Heterodera glycines* [65]. Isothiocyanates produced by members of the Brassicaceae family are toxic to *Tylenchulus semipenetrans* and inhibit the hatching of cysts and root-knot nematodes [66]. The secondary metabolite camalexin from *Arabidopsis* is known to inhibit both cyst and root-knot nematodes [67, 68]. Table **2** outlines some key secondary metabolites and their efficacy in preventing nematode infections.

Nematode Parasitism and Plant Immune Responses

In the course of coevolution, nematodes have developed sophisticated parasitic strategies to secure their survival and proliferation, which can be elucidated by examining the changes in their feeding sites within host root cells. These sites are characterized by extensive biochemical, molecular, and morphological alterations [101]. Various gene signaling pathways are suppressed by effector proteins secreted by nematodes at these feeding sites, thereby compromising the plant's innate immune defences [102, 103]. The only nematode hormone currently recognized, ascaroside Ascr#18, functions as a nematode-associated molecular pattern (NAMP). This molecule is detected by the plant immune system, triggering microbe-associated molecular pattern (MAMP)-triggered immunity (MTI) and pathogen-associated molecular pattern (PAMP)-triggered immunity (PTI), along with activating mitogen-activated protein kinase (MAPK) signalling and salicylic acid and jasmonic acid-mediated pathways in both root-knot and cyst nematodes [104]. Similar immune responses have been observed in tomato, potato, and barley plants. The PTI responses in host plants can be suppressed by various virulence effectors produced by both root-knot and cyst nematodes [105 - 107]. The leucine-rich repeat receptor-like kinase (LRR-RLK), encoded by the *Arabidopsis* gene *NILR1* (nematode-induced LRR-RLK 1), represents the first identified pattern recognition receptor (PRR) involved in PTI activation in response to ligands derived from *Heterodera schachtii* and *Meloidogyne incognita* [108]. The importance of PTI in nematode immunity has also been substantiated in *Arabidopsis* PTI-deficient mutants, where susceptibility was heightened in *bak1–5* and *bik1* mutants [67, 108]. Recent investigations have highlighted the role of nonribosomal peptide synthetase (NRPS)-like enzymes within the genomes of plant-parasitic nematodes (PPN). These enzymes share homology with fungal NRPS-like ATRR enzymes and are characterized by the presence of two sequential reducing domains at their N-terminus [A-T-R1-R2]. Experimental studies have examined the function of the NRPS-like enzyme MiATRR in the parasitism of *M. incognita*. Host-induced RNA interference (RNAi) targeting MiATRR in *A. thaliana* led to a marked reduction in gall formation, egg mass production, nematode growth rate, and overall body size [109].

Table 2. List of some of the important metabolites produced by plants against different plant parasitic nematodes.

Compound	Name of Enzyme/Gene/Metabolite	Accumulation Site	Function	Host Crops	Tested Against	References
Phenolic Compounds	Phenylalanine ammonia-lyase	Roots	When grown at 32°C, which inhibited enzyme activity, plants grown at 27°C exhibited optimal PAL activity, allowing the plants to synthesize phenylpropanoids used in nematode resistance.	Tomato, *Lycopersicum esculentum*	Root-knot nematode, *Meloidogyne incognita*	[69]
	a. Phenylalanine ammonia-lyase	Roots	Nematodes usually lead to higher levels of gene expression in resistant plants.	Alfalfa, *Medicago sativa*	Root-lesion nematode, *Pratylenchus penetrans*	[62]
	b. Chalcone synthase			Soybean, *Glycine max*	*Heterodera glycines* and *Meloidogyne incognita*	[70]
	c. Chalcone isomerase					
	d. Isoflavone reductase			Cowpea, *Vigna unguiculata* L. Walp	*Meloidogyne incognita*	[71]
	e. Caffeic acid O-methyltransferase					
	f. 4-coumarate-CoA ligase					
	g. Cinnamoyl CoA reductase			Soybean, *Glycine max* genotype PI 88788	Soybean cyst nematode, *Heterodera glycines* population NL1-RHg/HG-type 7	[72]
	h. Dihydroflavonol 4-reductase					
Sakuranetin	-	Leaf	Only identified among resistant cultivars; suggested to play a defensive role.	Rice, *Oryza sativa*	Stem nematode, *Ditylenchus angustus*	[57]

(Table 2) cont.....

Compound	Name of Enzyme/Gene/Metabolite	Accumulation Site	Function	Host Crops	Tested Against	References
Group 1	a. Formononetin and Formononetin-7-O-glu-coside-6"-O-malonate	Roots, meristems, leaves	Nematode resistance was associated with the buildup of isoflavonoids and pterocarpan (conjugate). *P. penetrans* was suppressed by medicarpin in a concentration-dependent way.	White clover, *Trifolium repens*	Stem nematode, *Ditylenchus dipsaci*	[73]
	b. Medicarpin-3-O-gluco-side-6"-O-malonate			Lucerne, *Medicago sativa*	Stem nematode, *Ditylenchus dipsaci*	[74]
	c. Medicarpin			Lucerne, *Medicago sativa*	Root-lesion nematode, *Pratylenchus penetrans*	[62]
	d. Coumesterol glucosides					
Group 2	a. O-methyl-apigenin- C-hexoside -O-deoxyhexoside	Roots and shoots during *P. neglectus* and *H. avenae* infection	Plants treated with methyl jasmonate and nematodes appeared to produce flavonoids, which may have functioned as broad defensive chemicals. Plants treated with root extracts from plants stimulated with methyl jasmonate exhibited decreased levels of infection.	Oats, *Avena sativa*	Root lesion nematode, *Pratylenchus neglectus*, Cereal cyst nematode, *Heteroderaavenae*, Stem nematode, *Ditylenchus dipsaci*	[75]
	b. Apigenin-C-hexoside-O -pentoside					
	c. Luteolin-C-hexoside-O					
Group 3	a. Coumesterol	Roots	In lima beans only, psoralidin and cholesterol collected in the roots and were localized in nematode-caused lesion sites. At dosages of 10–15 μg/mL, cholesterol greatly reduced the motility of nematodes.	Lima bean, *Phaseolus lunatus* and snap bean, *P. vulgaris*	Root-lesion nematode, *Pratylenchus scribneri*	[76]
	b. Psoralidin					
Group 4	a. Quercentagetin (hydroxy-flavone)	Adult female extracts	The flavonoid quercetagetin, which was present in pathotypes with a yellow colour and absent in pathotypes with a lighter hue, is responsible for the yellow colouration seen in *G. rostochiensis* and *G. pallida*.	-	Potato cyst nematodes, *Globodera rostochiensisand G. pallida*	[77]
	b. Aurone					
	c. Chalcone					
Group 5	a. Flavan-3,4-diols	Roots	Following nematode infection, condensed tannins and flavan-3, 4-diols accumulated.	Banana, *Musa* sp.	Burrowing nematode, *Radopholussimili*	[78]
	b. Condensed tannins					

(Table 2) cont.....

Compound	Name of Enzyme/Gene/Metabolite	Accumulation Site	Function	Host Crops	Tested Against	References
Group 6	a. Daidzein b. Genistein c. Other isoflavonoids	Roots	At two and four days after inoculation, susceptible Sussex cultivars showed increases in daidzein and genistein. At two and three days after inoculation, isoflavonoid production was increased in nematode-infected plants of the susceptible Sussex and resistant Hartwig cultivars.	Soybean, *Glycine max*	Soybean cyst nematode, *Heterodera glycines*	[79]
Glyceollin	-	Roots	In comparison to the susceptible cultivar, the resistant cultivar had higher accumulations of glyceollin I and III.	Soybean, *Glycine max*	Soybean cyst nematode, *Heterodera glycines*	[80]
		Stele in roots	Glyceollin was linked to a negative interaction between *M. incognita* and the resistant cultivar: Accumulation was seen in the resistant roots' stele, high glyceollin concentrations in the resistant cultivar, and glyceollin impeded motility in *M. incognita*.	Soybean, *Glycine max*	Root-knot nematodes, *Meloidogyne incognita* and *M. javanica*	[63]
		Leaves	At infection sites, glycocololin accumulated to high enough concentrations to cause a localized hypersensitivity reaction. Through the suppression of the mitochondrial electron transport pathway, it prevented nematode respiration and motility as well as plant tissue death.	*In vitro* system	Root-knot nematodes, *Meloidogyne incognita* and *M. javanica*	[63]
Phaseollin	-	Hypocotyl and root	Phaseollin is exclusive to tissue infected with *P. penetrans*. *P. penetrans* juveniles cultured in 47 µg/mL phaseollin solution for 16 hours did not lose any of their survival.	Common bean, *Phaseolus vulgaris*	Root-lesion nematode, *Pratylenchus penetrans*	[81]

(Table 2) cont.....

Compound	Name of Enzyme/Gene/Metabolite	Accumulation Site	Function	Host Crops	Tested Against	References
Flavonoid pathways	a. *Arabidopsis*: chalcone synthase, chalcone isomerase and flavonoid 3' hydroxylase, dihydroflavonol 4-reductase	Roots	Significantly more *M. incognita* reproduction was seen in tobacco mutants with increased anthocyanidin concentration. Reproduction of *M. incognita* was comparable in plants of the natural type and *Arabidopsis* tt mutants.	Tobacco, *Nicotiana tabacum* and *Arabidopsis thaliana*	Root-knot nematode, *M. incognita*	[60]
	b. Tobacco: Phenylalanine ammonia lyase, anthocyanidins					
	c. Quercetin 7-glucoside and other phenols	Root extracts	Root extracts decreased the number of galls, decreased nematode egg hatching, and inhibited nematode movement.	Common Lantana, *Lantana camara* L.	Root-knot nematode, *Meloidogyne incognita*	[82]
Chalcone isomerase Auxin-induced protein	-	Roots	Auxin-induced protein and chalcone isomerase protein.	Cowpea. *Vigna unguiculata* L. Walp	Root-knot nematode, *Meloidogyne incognita*	[83]
Group 1	a. Chalcone synthase	Root tissue	PIN 2 and flavonoid production genes are upregulated in nematode-infected roots.	Soybean, *Glycine max* L. Merr.	Soybean cyst nematode, *Heterodera glycines*	[82]
	b. Chalcone flavanone isomerase					
	c. Isoflavone reductase [putative]					
	d. Dihydroflavonol 4-reductase					
	e. Quercetin 3-O-methyltransferase					
	f. 4-coumarate-CoA ligase					
	g. PIN 2					
Group 2	a. Chalcone synthase genes, CHS1, CHS2, CHS3	Root tissue	CHS1: gusA, CHS2: gusA and CHS3: gusA expressions overlapped with GH3: gusA expression at 48 h, 72 h and 120 h post-inoculation.	White clover, *Trifolium repens*	Root-knot nematode, *Meloidogyne javanica*	[84]
	b. Auxin responsive gene, GH3					
Group 3	a. tt4 (chalcone synthase) mutant	N/A	The number of adult females in the tt (transparent testa) mutant lines of single and double tt4, tt5, and tt6 were not decreased by flavonoid deficiency; in fact, several lines produced more female nematodes.	*Arabidopsis thaliana*	Sugar beet nematode, *Heterodera schachtii*	[85]
	b. tt5 (chalcone isomerase) mutant					
	c. tt6 (flavonoid 3' hydroxylase) mutant					
Chalcone synthase (silencing by RNA interference)	-	Root tissue	The lack of flavonoids did not affect gall counts. In comparison to roots containing flavonoids, flavonoid-deficient roots had shorter galls and reduced pericycle cell division.	Barrel medic, *Medicago truncatula*	Root-knot nematode, *Meloidogyne javanica*	[86]

(Table 2) cont.....

Compound	Name of Enzyme/Gene/Metabolite	Accumulation Site	Function	Host Crops	Tested Against	References
Several compounds from the chalcone, flavone, flavanone, isoflavonoid and flavonol pathways	-	Purified compounds and plant extracts	At 60–84 µg/mL, kaempferol, quercetin, and myricetin repelled juvenile *M. incognita* and *R. similis*. At 100–142 µg/mL, luteolin, daidzein, and genistein all inhibited *R. similis* growth. After 48 hours of incubation, 13–41% of *M. incognita* juveniles were suppressed by kaempferol, quercetin, myricetin, rutin, and quercitrin. Up to 21% less eggs hatched in *R. similis* when naringenin and hesperetin, apigenin, daidzein, and kaempferol were present.	-	Burrowing nematode, *Radopholus similis*, root-lesion nematode, *Pratylenchus penetrans* and root-knot nematode, *Meloidogyne incognita*	[60]
	a. Patuletin	Purified compounds and marigold, *Tagetes patula* L. flower extracts	After 72 hours, patuletin killed all nematodes at all dilutions, while patulitrin killed between 10 and 50 per cent and quercetin between 70 and 80 per cent. Within 24 hours, rutin at 0.5–1% killed every nematode.	-	Corn cyst nematode, *Heterodera zeae*	[61]
	b. Patulitrin					
	c. Quercetin					
	d. Rutin					
(E)-chalcone	-	Purified compound	At the dose of 33 µM, (E)-chalcone killed nematodes in less than a day and prevented egg hatching.	N/A	Potato cyst nematodes, *Globodera rostochiensis* and *G. pallid*	[87]
Hydroxycinnamic Acids	-	Root	A sensitive poplar clone, *Populus tremula* × *Populus alba*, showed high repression of chlorogenic acid buildup in *Meloidogyne incognita* galls. This shows that *M. incognita* may use the suppression of chlorogenic acid production as a pathogenic strategy.	Tomato cultivar	Root knot nematode, *Meloidogyne incognita*	[88]
	a. (E)-cinnamic acid					
	b. Para-coumaric acid					
	c. Caffeic acid					

(Table 2) cont.....

Compound	Name of Enzyme/Gene/Metabolite	Accumulation Site	Function	Host Crops	Tested Against	References
Stilbenoids and Diarylheptanoids	-	bark and wood	Nematode resistance in grape vines and pine trees has been linked to stilbenoids.	Pine tree, *Pinus strobus*	pinewood nematode *Bursaphelenchus xylophilus*	[89]
	a. Phenylpropanoyl-CoA					
	b. C6-C2-C6 stilbene skeleton					
	c. 3-O-methyldihydropinosylvin					
	d. Phenylphenalenone phytoalexins	Root extracts	Three of the phenylphenalenones that were found in the extracts from the resistant banana cultivars demonstrated a noteworthy nematistatic activity in an *in vitro* experiment.	Banana cultivar	burrowing nematode *Radopholus similis*	[59]
Tannins	a. Flavan-3,4-diols	Root extracts	In addition to helping plants withstand insect herbivory, a few studies have discovered a link between tannin buildup and nematode resistance.	Banana cultivar	burrowing nematode *Radopholus similis*	[78]
	b. Flavan-3-ols			Pine tree, *Pinus strobus*	pinewood nematode *Bursaphelenchus xylophilus*	[90]
Terpenoids	a. Gossypol	Root extracts	With an IC50 of 10–50 µg/ml, a crude TA combination produced by extracting *G. hirsitum* roots and partially purifying it had substantial nematistatic activity toward *M. incognita* juveniles; doses as high as 125 µg/ml were likewise nematicidal. *A. thaliana* NPR1, a gene implicated in SA-mediated immunity, demonstrated increased resistance against many fungal diseases and the reniform worm *R. reniformis*.	Cotton, *Gossypium hirsitum*; pepper, *Capsicum annuum*	*Rotylenchulus reniformis*, Root knot nematode, *Meloidogyne incognita*	[91, 92]
	b. 6-O-methylgossypol					
	c. 6-O,6'-O dimethylgossypol					
Saponins	a. α-tomatine	Root	It has been investigated whether α-tomatine, which helps tomatoes resist fungal diseases and insect pests, can also help tomatoes resist *M. incognita*.	Tomato cultivar	Root knot nematode, *Meloidogyne incognita*	[93]
	b.α-solanine					
	c. α-chaconine	Tuber	According to one study, the high glycoalkaloid content of the wild potato species *Solanum canasense* may be responsible for its resistance to *G. pallida*.	Potato, *Solanum canasense*	Potato cyst nematode, *Globodera pallida*	[94]

(Table 2) cont.....

Compound	Name of Enzyme/Gene/Metabolite	Accumulation Site	Function	Host Crops	Tested Against	References
Alkaloids	a. Camalexin	-	When compared to the wild type, the cyp79b2/b3 double mutant, which has a large reduction in camalexin synthesis, demonstrated a statistically significant increase in *H. schachtii* reproduction. Similarly, *M. incognita* substantially more readily attacked the pad3 mutant, which is likewise deficient in the production of camalexin.	*Arabdopsis thaliana*	Sugarbeet nematode, *Heterodera schachtii*	[68]
	b. Pyrrolizidine	Root	Soil amendments made of plants rich in pyrrolizidine alkaloids decrease PPN, while pure pyrrolizidine alkaloid standards and extracts from pyrrolizidine alkaloid-rich plants are typically nematicidal *in vitro*.	Crotalaria	Different PPNs	[95]
	c. Ergot alkaloids (ergovaline)	Root	Based on the findings that ergot alkaloids are nematicidal *in vitro* and that grasses colonized by endophytic fungi, such as *Epichloë* spp., reduce Pratylenchus in field settings, ergot alkaloids may be involved in resistance to *Pratylenchus* sp.	Perennial ryegrass, *Lolium perenne*	Lesion nematodes, *Pratylenchus* sp.	[96]
Benzoxazinoids	a. HMBOA-glucoside	Root exudates	Wheat that was suppressed from producing benzoxazinoid by an arbuscular mycorrhizal fungus was far more vulnerable to *P. neglectus*.	Wheat, *Triticum aestivum*; rye, *Secale cereale*, and maize	Stubby-root nematode, *Paratrichodorus neglectus*	[97]
	b. HDMBOA-glucoside	Root exudates				

(Table 2) cont.....

Compound	Name of Enzyme/Gene/Metabolite	Accumulation Site	Function	Host Crops	Tested Against	References
Glucosinolates	a. Glucosinolate	-	The strong nematicidal action of glucosinolates is suggested by *in vitro* evidence, and biofumigation—the application of green manures or *Brassica* seed meal as soil amendments-can be a successful substitute for chemical fumigation.	*Brassica napus*	Stubby-root nematode, *Paratrichodorus neglectus*	[98]
	b. 2-phenylethyl glucosinolate					
	c. 4-methoxy-3-indolylmethyl glucosinolate	-	Moreover, large cells of *A. thaliana* infected with *M. incognita* exhibit a considerable downregulation of MYB34 expression, adding evidence to the possibility that glucosinolates contribute to PPN resistance.	*Arabidopsis thaliana*	Root knot nematode, *Meloidogyne incognita*	[99]
Organosulfur Compounds	a. α-terthienyl	Roots and their exudates	α-terthienyl owes its nematicidal effect to its ability to induce oxidative stress inside the nematode	Marigolds, *Tagetes* sp.	Lesion nematodes, *Pratylenchus penetrans*	[100]
	b. polythienyl					
	c. Asparagusic acid	-	The nematicidal impact of α-terthienyl is attributed to its capacity to cause oxidative stress within the worm. At 50 ppm, asparagusic acid exhibits potent nematicidal effects against multiple PPN species and prevents the hatching of *Heterodera* eggs.	Asparagus, *Asparagus officinalis*	*Heteroodera* sp.	[65]

Effector Molecules and Receptors in Plant-Nematode Interactions

Effector molecules are secreted by nematodes as a defensive mechanism, with various nematode organs such as the oesophagal gland, hypodermis, amphids, and phasmids playing pivotal roles in their secretion [109]. These effector molecules are instrumental in modulating feeding sites within plants, particularly those secreted by root-knot nematodes and cyst nematodes [110]. Certain proteins have been identified as effector molecules in the root-knot nematode (*Meloidogyne* spp.), including receptor-like kinases and receptor-like proteins that initiate immune responses in tomato plants, as reported by El-Sappah [111]. Critical immune responses demarcated during the early stages of nematode infestation include the production of reactive oxygen species (ROS), synthesis of secondary metabolites, cell death along the nematode's migratory pathway, and the reinforcement of plant cell walls [112]. Cell wall-degrading enzymes such as chitinases, pectinases, cellulases, endoglucanases, and xylanases, encoded by specific genes, facilitate nematode migration by softening the plant root tissue [113 - 116]. Certain cyst nematodes inject effector molecules such as GrVAP1 [117], RHA1B [118], Ha18764 [119], and GrCEP12 [107] into the host plant using stylets, synthesizing these molecules within their oesophagal gland cells [120]. The suppression of basal immune responses by cyst-forming nematodes enables the successful establishment of feeding relationships with susceptible host plants, promoting nematode growth and reproduction [101, 120]. Effector molecules Mi-EFF1 and Mj-NULG1a, secreted by the dorsal gland of root-knot nematodes (*Meloidogyne* spp.), have been shown to localize in the nuclei of giant cells in tomato plants, indicating their involvement in early nematode-plant interactions [121, 122].

HORMONES IN PLANT-NEMATODE INTERACTIONS

Plant hormones, synthesized in specific locations and subsequently distributed throughout the plant, are crucial in regulating various physiological and cellular functions, including stress responses, organogenesis, growth, metabolism, seed dormancy, and reproduction. Key plant hormones such as auxins, gibberellins, cytokinins, abscisic acid, and ethylene play significant roles in these processes. Studies on the root-knot nematode *Meloidogyne incognita* and the cyst nematode *Heterodera schachtii* have demonstrated that upregulation and downregulation of genes involved in auxin biosynthesis are critical for establishing nematode feeding sites in host plants. Upon nematode infection, genes promoting auxin biosynthesis become upregulated, leading to the accumulation of auxin at the feeding sites [123 - 125]. The role of auxin transport in nematode infection has also been validated in *Arabidopsis thaliana* [126]. Effector molecules 19C07 and 10A07 secreted by *H. schachtii* interact with auxin-related proteins, LAX3 and Indoleacetic Acid-Induced16 (IAA16), in *Arabidopsis*, enhancing auxin influx in syncytia [127, 103]. Cytokinins have similarly been implicated in the development of nematode-feeding sites. Research has identified nematode genes in *H. schachtii* and *M. incognita* that trigger cytokinin production [128]. Siddique *et al.* [129] demonstrated that silencing these genes in *H. schachtii* or inducing cytokinin biosynthesis mutations in *Arabidopsis* leads to the formation of smaller syncytia, suggesting that both nematode and plant-derived cytokinins are essential for feeding site development [129].

Research has yielded conflicting results regarding the role of ethylene, a gaseous hormone with diverse functions in plant biology. Ethylene has been found to inhibit root-knot nematode infection by reducing nematode attraction to the roots

[130], whereas it appears to enhance attraction in cyst nematodes [131]. Subsequent studies have shown that soybean cyst nematode (*Heterodera glycines*) juveniles exhibit increased attraction to plant roots when treated with ethylene inhibitors [132]. Recent findings indicate that ethylene signalling is associated with two distinct pathways [133], suggesting that depending on the specific host-nematode interaction and the timing or location of ethylene action, cross-talk with other hormonal pathways may yield varying effects on host response to cyst nematode infection. One pathway involves the suppression of salicylic acid-based defences, which increases plant susceptibility to cyst nematodes, explaining why ethylene enhances cyst nematode infection. Conversely, ethylene receptor 1-mediated inhibition of cytokinin signalling reduces susceptibility to cyst nematode infection. Both salicylic acid and gibberellic acid have been identified as defensive molecules against nematode infections in several studies [134].

CONCLUSION

The presence of proteins, neurotransmitters, and signalling mechanisms essential for environmental responsiveness is a common feature among all living organisms, ranging from simple viruses and bacteria to complex plants and animals. Consequently, it is not unexpected that plants produce signaling molecules to counteract nematode or parasite proteins. Given that the nematode's chemosensory signalling components interact directly with the external environment, they are susceptible to antagonistic chemicals that might inhibit their interaction with host stimuli. These signalling components are secreted from specialized glands. Plant parasitic nematodes cause substantial agricultural losses globally each year, and current strategies for managing these pests remain limited, despite the growing severity of nematode-related agricultural issues. Recent advances in understanding plant-nematode interactions have provided valuable insights. Innovations in technology are facilitating the discovery of effector compounds secreted by plant parasitic nematodes and phytochemicals by plants that prevent nematode penetration. Identifying the genes responsible for nematode parasitism and other nematode-specific functions will be crucial for developing new, durable plant defense strategies. An in-depth understanding of the interaction between plants and nematodes could pave the way for innovative control methods, as bridging the gap between fundamental research and practical applications becomes increasingly challenging.

ACKNOWLEDGEMENTS

The authors are thankful to the Vice Chancellor and Head of the, Department of Bioscience and Biotechnology, Banasthali Vidyapith, Rajasthan, for providing the necessary facilities to compile the work. Indian Council of Medical Research,

New Delhi is also acknowledged for providing financial assistance to authors through research project IIRP-2023-0918.

REFERENCES

[1]   Kranz J. Interactions in pest complexes and their effects on yield. J Plant Dis Prot 2005; 112: 366-85.

[2]   Döring TF, Pautasso M, Finckh MR, Wolfe MS. Concepts of plant health – reviewing and challenging the foundations of plant protection. Plant Pathol 2012; 61(1): 1-15.
[http://dx.doi.org/10.1111/j.1365-3059.2011.02501.x]

[3]   Oka Y. From old-generation to next-generation nematicides. Agronomy (Basel) 2020; 10(9): 1387.
[http://dx.doi.org/10.3390/agronomy10091387]

[4]   Hodda M. Phylum Nematoda: a classification, catalogue and index of valid genera, with a census of valid species. Zootaxa 2022; 5114(1): 1-289.
[http://dx.doi.org/10.11646/zootaxa.5114.1.1] [PMID: 35391386]

[5]   Palomares-Rius JE, Escobar C, Cabrera J, Vovlas A, Castillo P. Anatomical alterations in plant tissues induced by plant-parasitic nematodes. Front plant sci 2017; 8: 1987.
[http://dx.doi.org/10.3389/fpls.2017.01987]

[6]   Nicol JM, Turner SJ, Coyne DL, den Nijis L, Hockland S, Maafi ZT. Current nematode threats to world agriculture. In: Jones J, Gheysen G, Fenoll C, Eds. Genomics and molecular genetics of plant-nematode interactions. Dordrecht: Springer 2011; pp. 21-43.
[http://dx.doi.org/10.1007/978-94-007-0434-3_2]

[7]   Jones JT, Haegeman A, Danchin EGJ, *et al.* Top 10 plant-parasitic nematodes in molecular plant pathology. Mol Plant Pathol 2013; 14(9): 946-61.
[http://dx.doi.org/10.1111/mpp.12057] [PMID: 23809086]

[8]   Abd-Elgawad MMM, Askary TH. Impact of phytonematodes on agriculture economy. In: Askary TH, Martinelli PRP, Eds. Biocontrol agents of phytonematodes. CAB International 2015; pp. 3-49.
[http://dx.doi.org/10.1079/9781780643755.0003]

[9]   Grabau ZJ, Noling JW, Clark C, Overstreet C. Nematode management in sweet potatoes (including boniatos): ENY-030/NG030, 10/2021. EDIS. 2021;2021(5).
[http://dx.doi.org/10.32473/edis-ng030-2021]

[10]   Quick RA, Cimrhakl L, Mojtahedi H, Sathuvalli V, Feldman MJ, Brown CR. Elimination of *Tobacco rattle* virus from viruliferous *Paratrichodorus allius* in greenhouse pot experiments through cultivation of castle russet. J Nematol 2020; 52(1): 1-10.
[http://dx.doi.org/10.21307/jofnem-2020-011] [PMID: 32193908]

[11]   Thiessen L. Stubby Root Nematode of Soybean. NCState Extension Publications; 2020.

[12]   Aryal SK, Davis RF, Stevenson KL, Timper P, Ji P. Influence of infection of cotton by *Rotylenchulus reniformis* and *Meloidogyne incognita* on the production of enzymes involved in systemic acquired resistance. J Nematol 2011; 43(3-4): 152-9.
[PMID: 23431029]

[13]   Eisenback JD, Hartman KM. *Sphaeronema sasseri* n. sp. (Tylenchulidae), a nematode parasitic on fraser fir and red spruce. J Nematol 1985; 17(3): 346-54.
[PMID: 19294105]

[14]   McSorley R, Parrado JL. The spiral nematode *Helicotylenchus multicinctus* on bananas in Florida and its control. Proc Annu Meet Fla State Hort Soc 1983; 96: 201-7.

[15]   Baines RC, Miyakawa T, Cameron JW, Small RH. Infectivity of two biotypes of the citrus nematode on citrus and on some other hosts. J Nematol 1969; 1(2): 150-9.
[PMID: 19325669]

[16]   Inserra RN, Duncan LW, O'Bannon JH, Fuller SA. Citrus nematode biotypes and resistant citrus rootstocks in Florida Nematol Circular No 205 Florida Department of Agriculture and Consumer Services. Division of Plant Industry 1994.

[17]   Davis EL, MacGuidwin AE. Lesion nematode disease. Plant Health Instr 2000.
[http://dx.doi.org/10.1094/PHI-I-2000-1030-02]

[18]   Keshari AK, Gupta R. Five species of plant parasitic nematodes (Order: Tylenchidae) affecting vegetable crops of Nepal. Int J Life Sci 2016; 10(1): 10-4.
[http://dx.doi.org/10.3126/ijls.v10i1.14515]

[19]   Brooks FE. BurrowingNematode. Plant Health Instructor 2008; 08.
[http://dx.doi.org/10.1094/PHI-I-2008-1020-01]

[20]   Trivino R, Cano JL, Jamieson C. An innovative agroforestry system for food and fuel production. J Environ Sci Manag 2016; 1: 74-82.
[http://dx.doi.org/10.47125/jesam/2016_sp1/06]

[21]   Ibrahim IKA, Mokbel AA, Handoo ZA. Current status of phytoparasitic nematodes and their host plants in Egypt. Nematropica 2010; 40: 239-62.

[22]   Mitkowski NA, Abawi GS. Root-knot Nematode. Plant Health Instruct 2003; 03
[http://dx.doi.org/0.1094/PHI-I-2003-0917-01]

[23]   Di Vito M. The Pea cyst nematode, *Heterodera goettingiana*. Nematology circular no. 188; 1991.

[24]   Mugniéry D, Phillips MS. The Nematode Parasites of Potato. In: Vreugdenhil D, Bradshaw J, Gebhardt C, Govers F, Taylor MA, MacKerron DKL, Ross HA Eds., Potato biology and biotechnology, advances and perspectives, 2007,569-594.

[25]   Donald PA, Stamps WT, Linit MJ, Todd TC. Pine wilt. Plant Health Instr 2003.
[http://dx.doi.org/10.1094/PHI-I-2003-0130-01]

[26]   Zhang Y, Wen TY, Wu XQ, Hu LJ, Qiu YJ, Rui L. The *Bursaphelenchus xylophilus* effector BxML1 targets the cyclophilin protein (CyP) to promote parasitism and virulence in pine. BMC Plant Biol 2022; 22(1): 216.
[http://dx.doi.org/10.1186/s12870-022-03567-z] [PMID: 35473472]

[27]   Franklin MT, Hooper DJ. *Bursaphelenchus fungivorus* n. sp. (Nematoda: Aphelenchoidea) from rotting gardenia buds infected with *Botrytis cinerea*. Nematologica 1962; 8(2): 136-42.
[http://dx.doi.org/10.1163/187529262X00350]

[28]   Tulek A, Kepenekci I, Cifticigil TH, *et al.* Effects of seed-gall nematode, *Anguina tritici,* on bread wheat grain characteristics and yields in Turkey. Nematology 2015; 17(9): 1099-104.
[http://dx.doi.org/10.1163/15685411-00002926]

[29]   Dash M, Somvanshi VS, Walia RK. Stem and bulb nematodes in agricultural crops and their management by biological and biotechnological methods. In: Khan MR, Ed Novel Biological and Biotechnological Applications in Plant Nematode Management. Singapore: Springer Nature Singapore 2023; 9: pp. 341-58.
[http://dx.doi.org/10.1007/978-981-99-2893-4_15]

[30]   Khan MR, Handoo ZA, Rao U, Rao SB, Prasad JS. Observations on the Foliar Nematode, *Aphelenchoides besseyi,* infecting tuberose and rice in india. J Nematol 2012; 44(4): 391-8.
[PMID: 23482906]

[31]   Jagdale GB, Grewal PS. Identification of alternatives for the management of foliar nematodes in floriculture. Pest Manag Sci 2002; 58(5): 451-8.
[http://dx.doi.org/10.1002/ps.472] [PMID: 11997971]

[32]   Mitchum MG, Liu X. Peptide effectors in phytonematode parasitism and beyond. Annu Rev Phytopathol 2022; 60(1): 97-119.
[http://dx.doi.org/10.1146/annurev-phyto-021621-115932] [PMID: 35385672]

[33] Wang M, Weiberg A, Jin H. Pathogen small RNAS : a new class of effectors for pathogen attacks. Mol Plant Pathol 2015; 16(3): 219-23.
[http://dx.doi.org/10.1111/mpp.12233] [PMID: 25764211]

[34] Rangel LI, Bolton MD. The unsung roles of microbial secondary metabolite effectors in the plant disease cacophony. Curr Opin Plant Biol 2022; 68: 102233.
[http://dx.doi.org/10.1016/j.pbi.2022.102233] [PMID: 35679804]

[35] Hussey RS, Mims CW. Ultrastructure of esophageal glands and their secretory granules in the root-knot nematode *Meloidogyne incognita.* Protoplasma 1990; 156(1-2): 9-18.
[http://dx.doi.org/10.1007/BF01666501]

[36] Davis EL, Hussey RS, Mitchum MG, Baum TJ. Parasitism proteins in nematode–plant interactions. Curr Opin Plant Biol 2008; 11(4): 360-6.
[http://dx.doi.org/10.1016/j.pbi.2008.04.003] [PMID: 18499507]

[37] Fosu-Nyarko J, Jones MGK. Advances in understanding the molecular mechanisms of root lesion nematode host interactions. Annu Rev Phytopathol 2016; 54: 253-78.
[http://dx.doi.org/10.1146/annurev-phyto-080615-100257] [PMID: 27296144]

[38] Grundler FMW, Sobczak M, Golinowski W. Formation of wall openings in root cells of *Arabidopsis thaliana* following infection by the plant-parasitic nematode *Heterodera schachtii.* Eur J Plant Pathol 1998; 104(6): 545-51.
[http://dx.doi.org/10.1023/A:1008692022279]

[39] Golinowski W, Grundler FMW, Sobczak M. Changes in the structure of *Arabidopsis thaliana* during female development of the plant-parasitic nematode *Heterodera schachtii.* Protoplasma 1996; 194(1-2): 103-16.
[http://dx.doi.org/10.1007/BF01273172]

[40] Jones MG, Payne HL. Early stages of nematode-induced giant-cell formation in roots of *Impatiens balsamina.* J Nematol 1978; 10(1): 70-84.
[PMID: 19305816]

[41] Anjam MS, Shah SJ, Matera C, *et al.* Host factors influence the sex of nematodes parasitizing roots of *Arabidopsis thaliana.* Plant Cell Environ 2020; 43(5): 1160-74.
[http://dx.doi.org/10.1111/pce.13728] [PMID: 32103526]

[42] Ellenby C. Environmental determination of the sex ratio of a plant parasitic nematode. Nature 1954; 174(4439): 1016-7.
[http://dx.doi.org/10.1038/1741016b0]

[43] Engelbrecht G, Claassens S, Mienie CMS, Fourie H. South Africa: an important soybean producer in sub-Saharan Africa and the quest for managing nematode pests of the crop. Agriculture 2020; 10(6): 242.
[http://dx.doi.org/10.3390/agriculture10060242]

[44] Abd-Elgawad MMM. Understanding molecular plant–nematode interactions to develop alternative approaches for nematode control. Plants 2022; 11(16): 2141.
[http://dx.doi.org/10.3390/plants11162141] [PMID: 36015444]

[45] Dodds PN, Rathjen JP. Plant immunity: towards an integrated view of plant–pathogen interactions. Nat Rev Genet 2010; 11(8): 539-48.
[http://dx.doi.org/10.1038/nrg2812] [PMID: 20585331]

[46] Hartmann T. From waste products to ecochemicals: Fifty years research of plant secondary metabolism. Phytochemistry 2007; 68(22-24): 2831-46.
[http://dx.doi.org/10.1016/j.phytochem.2007.09.017] [PMID: 17980895]

[47] Seigler DS. Plant secondary metabolism. 1st ed. New York: Springer Science+Business Media; 1998.
[http://dx.doi.org/10.1007/978-1-4615-4913-0]

[48]   Viant MR, Kurland IJ, Jones MR, Dunn WB. How close are we to complete annotation of metabolomes? Curr Opin Chem Biol 2017; 36: 64-9.
[http://dx.doi.org/10.1016/j.cbpa.2017.01.001] [PMID: 28113135]

[49]   Sharma BR, Kumar V, Gat Y, Kumar N, Parashar A, Pinakin DJ. Microbial maceration: a sustainable approach for phytochemical extraction. 3 Biotech 2018; 8(9): 401.
[http://dx.doi.org/10.1007/s13205-018-1423-8] [PMID: 30221114]

[50]   Jaeger R, Cuny E. Terpenoids with special pharmacological significance: A review. Nat Prod Commun. 2016; 11.
[http://dx.doi.org/10.1177/1934578X1601100946]

[51]   Osbourn AE. Preformed antimicrobial compounds and plant defense against fungal attack. Plant Cell 1996; 8(10): 1821-31.
[http://dx.doi.org/10.2307/3870232] [PMID: 12239364]

[52]   VanEtten HD, Mansfield JW, Bailey JA, Farmer EE. Two classes of plant antibiotics: phytoalexins versus "phytoanticipins". Plant Cell 1994; 6(9): 1191-2.
[http://dx.doi.org/10.2307/3869817] [PMID: 12244269]

[53]   Donnez D, Jeandet P, Clément C, Courot E. Bioproduction of resveratrol and stilbene derivatives by plant cells and microorganisms. Trends Biotechnol 2009; 27(12): 706-13.
[http://dx.doi.org/10.1016/j.tibtech.2009.09.005] [PMID: 19875185]

[54]   Fry FH, Okarter N, Baynton-Smith C, *et al.* Use of a substrate/alliinase combination to generate antifungal activity in situ. J Agric Food Chem 2005; 53(3): 574-80.
[http://dx.doi.org/10.1021/jf048481j] [PMID: 15686404]

[55]   Virtanen AI, Matikkala EJ. Isolation of S-methyl and S-propyl cysteine sulfoxide from onion and antibiotic activity of crushed onion. Acta Chem Scand 1959; 13: 1898-900.
[http://dx.doi.org/10.3891/acta.chem.scand.13-1898]

[56]   Pegard A, Brizzard G, Fazari A, Soucaze O, Abad P, Djian-Caporalino C. Histological characterization of resistance to different root-knot nematode species related to phenolics accumulation in *Capsicum annuum*. Phytopathology 2005; 95(2): 158-65.
[http://dx.doi.org/10.1094/PHYTO-95-0158] [PMID: 18943985]

[57]   Gill JR, Harbornez JB, Plowright RA, Grayer RJ, Rahman ML. The induction of phenolic compounds in rice after infection by the stem nematode *Ditylenchus angustus*. Nematologica 1996; 42(5): 564-78.
[http://dx.doi.org/10.1163/004625996X00063]

[58]   D'Addabbo T, Carbonara T, Argentieri MP, *et al.* Nematicidal potential of *Artemisia annua* and its main metabolites. Eur J Plant Pathol 2013; 137(2): 295-304.
[http://dx.doi.org/10.1007/s10658-013-0240-5]

[59]   Hölscher D, Dhakshinamoorthy S, Alexandrov T, *et al.* Phenalenone-type phytoalexins mediate resistance of banana plants (*Musa* spp.) to the burrowing nematode *Radopholus similis*. Proc Natl Acad Sci USA 2014; 111(1): 105-10.
[http://dx.doi.org/10.1073/pnas.1314168110] [PMID: 24324151]

[59]   Hu Y, You J, Li C, Williamson VM, Wang C. Ethylene response pathway modulates attractiveness of plant roots to soybean cyst nematode *Heterodera glycines*. Sci Rep 2017; 7(1): 41282.
[http://dx.doi.org/10.1038/srep41282] [PMID: 28112257]

[60]   Chin S, Behm CA, Mathesius U. Functions of flavonoids in plant–nematode interactions. Plants 2018; 7(4): 85.
[http://dx.doi.org/10.3390/plants7040085] [PMID: 30326617]

[61]   Faizi S, Fayyaz S, Bano S, *et al.* Isolation of nematicidal compounds from *Tagetes patula* L. yellow flowers: structure-activity relationship studies against cyst nematode *Heterodera zeae* infective stage larvae. J Agric Food Chem 2011; 59(17): 9080-93.
[http://dx.doi.org/10.1021/jf201611b] [PMID: 21780738]

[62]   Baldridge GD, O'Neill NR, Samac DA. Alfalfa (*Medicago sativa* L.) resistance to the root-lesion nematode, *Pratylenchus penetrans*: defense-response gene mRNA and isoflavonoid phytoalexin levels in roots. Plant Mol Biol 1998; 38(6): 999-1010.
[http://dx.doi.org/10.1023/A:1006182908528] [PMID: 9869406]

[63]   Kaplan DT, Keen NT, Thomason IJ. Association of glyceollin with the incompatible response of soybean roots to *Meloidogyne incognita.* Physiol Plant Pathol 1980; 16(3): 309-18.
[http://dx.doi.org/10.1016/S0048-4059(80)80002-6]

[64]   Hamaguchi T, Sato K, Vicente CSL, Hasegawa K. Nematicidal actions of the marigold exudate α-terthienyl: oxidative stress-inducing compound penetrates nematode hypodermis. Biol Open 2019; 8(4): bio.038646.
[http://dx.doi.org/10.1242/bio.038646] [PMID: 30926596]

[65]   Takasugi M, Yachida Y, Anetai M, Masamune T, Kegasawa K. Identification of asparagusic acid as a nematicide occurring naturally in the roots of asparagus. Chem Lett 1975; 4(1): 43-4.
[http://dx.doi.org/10.1246/cl.1975.43]

[66]   Zasada IA, Meyer SLF, Halbrendt JM, Rice C. Activity of hydroxamic acids from *Secale cereale* against the plant-parasitic nematodes *Meloidogyne incognita* and *Xiphinema americanum.* Phytopathology 2005; 95(10): 1116-21.
[http://dx.doi.org/10.1094/PHYTO-95-1116] [PMID: 18943462]

[67]   Teixeira MA, Wei L, Kaloshian I. Root-knot nematodes induce pattern-triggered immunity in *Arabidopsis thaliana* roots. New Phytol 2016; 211(1): 276-87.
[http://dx.doi.org/10.1111/nph.13893] [PMID: 26892116]

[68]   Shah SJ, Anjam MS, Mendy B, *et al.* Damage-associated responses of the host contribute to defence against cyst nematodes but not root-knot nematodes. J Exp Bot 2017; 68(21-22): 5949-60.
[http://dx.doi.org/10.1093/jxb/erx374] [PMID: 29053864]

[69]   Brueske CH. Phenylalanine ammonia lyase activity in tomato roots infected and resistant to the root-knot nematode, *Meloidogyne incognita.* Physiol Plant Pathol 1980; 16(3): 409-14.
[http://dx.doi.org/10.1016/S0048-4059(80)80012-9]

[70]   Edens RM, Anand SC, Bolla RI. Enzymes of the phenylpropanoid pathway in soybean infected with *Meloidogyne incognita* or *Heterodera glycines.* J Nematol 1995; 27(3): 292-303.
[PMID: 19277292]

[71]   Villeth GRC, Carmo LST, Silva LP, *et al.* Cowpea– *Meloidogyne incognita* interaction: Root proteomic analysis during early stages of nematode infection. Proteomics 2015; 15(10): 1746-59.
[http://dx.doi.org/10.1002/pmic.201400561] [PMID: 25736976]

[72]   Klink VP, Lawrence GW, Matsye PD, Showmaker KC. The application of a developmental genomics approach to study the resistant reaction of soybean to the soybean cyst nematode. Nematropica 2010; 40(1): 1-11.

[73]   Cook R, Tiller SA, Mizen KA, Edwards R. Isoflavonoid metabolism in resistant and susceptible cultivars of white clover infected with the stem nematode *Ditylenchus dipsaci.* J Plant Physiol 1995; 146(3): 348-54.
[http://dx.doi.org/10.1016/S0176-1617(11)82067-5]

[74]   Edwards R, Mizen T, Cook R. Isoflavonoid conjugate accumulation in the roots of lucerne (*Medicago sativa*) seedlings following infection by the stem nematode (*Ditylenchus dipsaci*). Nematologica 1995; 41(1-4): 51-66.
[http://dx.doi.org/10.1163/003925995X00044]

[75]   Soriano IR, Asenstorfer RE, Schmidt O, Riley IT. Inducible flavone in oats (*Avena sativa*) is a novel defense against plant-parasitic nematodes. Phytopathology 2004; 94(11): 1207-14.
[http://dx.doi.org/10.1094/PHYTO.2004.94.11.1207] [PMID: 18944456]

[76]   Rich JR, Keen NT, Thomason IJ. Association of coumestanss with tee hypersensitivity of Lima bean

roots to Pratylenchus scribneri. Physiol Plant Pathol 1977; 10(2): 105-16.
[http://dx.doi.org/10.1016/0048-4059(77)90014-5]

[77]   Vlachopoulos EG, Smith L. Flavonoids in potato cyst nematodes. Fundam Appl Nematol 1993; 16: 103-6.

[78]   Collingborn FMB, Gowen SR, Mueller-Harvey I. Investigations into the biochemical basis for nematode resistance in roots of three musa cultivars in response to *Radopholus similis* infection. J Agric Food Chem 2000; 48(11): 5297-301.
[http://dx.doi.org/10.1021/jf000492z] [PMID: 11087475]

[79]   Kennedy MJ, Niblack TL, Krishnan HB. Infection by *Heterodera glycines* elevates isoflavonoid production and influences soybean nodulation. J Nematol 1999; 31(3): 341-7.
[PMID: 19270906]

[80]   Huang JS, Barker KR. Glyceollin I in soybean-cyst nematode interactions : spatial and temporal distribution in roots of resistant and susceptible soybeans. Plant Physiol 1991; 96(4): 1302-7.
[http://dx.doi.org/10.1104/pp.96.4.1302] [PMID: 16668334]

[81]   Abawi GS, Van Etten HD, Mai WF. Phaseollin production induced by *Pratylenchus penetrans* in *Phaseolus vulgaris*. J Nematol 1971; 3: 301.

[82]   Ithal N, Recknor J, Nettleton D, *et al.* Parallel genome-wide expression profiling of host and pathogen during soybean cyst nematode infection of soybean. Mol Plant Microbe Interact 2007; 20(3): 293-305.
[http://dx.doi.org/10.1094/MPMI-20-3-0293] [PMID: 17378432]

[83]   Oliveira J, Araujo-Filho J, Grangeiro T, *et al.* Enhanced synthesis of antioxidant enzymes, defense proteins and leghemoglobin in rhizobium-free cowpea roots after challenging with *Meloidogyne incognita*. Proteomes 2014; 2(4): 527-49.
[http://dx.doi.org/10.3390/proteomes2040527] [PMID: 28250394]

[84]   Hutangura P, Mathesius U, Jones MGK, Rolfe BG. Auxin induction is a trigger for root gall formation caused by root-knot nematodes in white clover and is associated with the activation of the flavonoid pathway. Funct Plant Biol 1999; 26(3): 221-31.
[http://dx.doi.org/10.1071/PP98157]

[85]   Jones JT, Furlanetto C, Phillips MS. The role of flavonoids produced in response to cyst nematode infection of *Arabidopsis thaliana*. Nematology 2007; 9(5): 671-7.
[http://dx.doi.org/10.1163/156854107782024875]

[86]   Wasson AP, Ramsay K, Jones MGK, Mathesius U. Differing requirements for flavonoids during the formation of lateral roots, nodules and root knot nematode galls in *Medicago truncatula*. New Phytol 2009; 183(1): 167-79.
[http://dx.doi.org/10.1111/j.1469-8137.2009.02850.x] [PMID: 19402878]

[87]   González JA, Estévez-Braun A. Effect of (*E*)-Chalcone on potato-cyst nematodes (*Globodera pallida* and *G. rostochiensis*). J Agric Food Chem 1998; 46(3): 1163-5.
[http://dx.doi.org/10.1021/jf9706686]

[88]   Baldacci-Cresp F, Behr M, Kohler A, *et al.* Molecular changes concomitant with vascular system development in mature galls induced by root-knot nematodes in the model tree host *Populus tremula* × *P. alba*. Int J Mol Sci 2020; 21(2): 406.
[http://dx.doi.org/10.3390/ijms21020406] [PMID: 31936440]

[89]   Hanawa F, Yamada T, Nakashima T. Phytoalexins from *Pinus strobus* bark infected with pinewood nematode, *Bursaphelenchus xylophilus*. Phytochemistry 2001; 57(2): 223-8.
[http://dx.doi.org/10.1016/S0031-9422(00)00514-8] [PMID: 11382237]

[90]   Pimentel CS, Firmino PN, Calvão T, Ayres MP, Miranda I, Pereira H. Pinewood nematode population growth in relation to pine phloem chemical composition. Plant Pathol 2017; 66(5): 856-64.
[http://dx.doi.org/10.1111/ppa.12638]

[91]   Veech JA. Histochemical localization and nematoxicity of terpenoid aldehydes in cotton. J Nematol

1979; 11(3): 240-6.
[PMID: 19300641]

[92] Parkhi V, Kumar V, Campbell LM, Bell AA, Shah J, Rathore KS. Resistance against various fungal pathogens and reniform nematode in transgenic cotton plants expressing *Arabidopsis* NPR1. Transgenic Res 2010; 19(6): 959-75.
[http://dx.doi.org/10.1007/s11248-010-9374-9] [PMID: 20151323]

[93] Elliger CA, Waiss AC Jr, Dutton HL, Rose MF. α-Tomatine and resistance of tomato cultivars toward the nematode,*Meloidogyne incognita.* J Chem Ecol 1988; 14(4): 1253-9.
[http://dx.doi.org/10.1007/BF01019350] [PMID: 24276208]

[94] Castelli L, Bryan G, Blok VG, *et al.* Investigations of *Globodera pallida* invasion and syncytia formation within roots of the susceptible potato cultivar Désirée and resistant species *Solanum canasense.* Nematology 2006; 8(1): 103-10.
[http://dx.doi.org/10.1163/156854106776180005]

[95] Thoden TC, Boppré M, Hallmann J. Effects of pyrrolizidine alkaloids on the performance of plant-parasitic and free-living nematodes. Pest Manag Sci 2009; 65(7): 823-30.
[http://dx.doi.org/10.1002/ps.1764] [PMID: 19378265]

[96] Bacetty AA, Snook ME, Glenn AE, *et al.* Toxicity of endophyte-infected tall fescue alkaloids and grass metabolites on *Pratylenchus scribneri.* Phytopathology 2009; 99(12): 1336-45.
[http://dx.doi.org/10.1094/PHYTO-99-12-1336] [PMID: 19899999]

[97] Frew A, Powell JR, Glauser G, Bennett AE, Johnson SN. Mycorrhizal fungi enhance nutrient uptake but disarm defences in plant roots, promoting plant-parasitic nematode populations. Soil Biol Biochem 2018; 126: 123-32.
[http://dx.doi.org/10.1016/j.soilbio.2018.08.019]

[98] Potter MJ, Vanstone VA, Davies KA, Kirkegaard JA, Rathjen AJ. Reduced susceptibility of *Brassica napus* to *Pratylenchus neglectus* in plants with elevated root levels of 2-phenylethyl glucosinolate. J Nematol 1999; 31(3): 291-8.
[PMID: 19270899]

[99] Portillo M, Cabrera J, Lindsey K, *et al.* Distinct and conserved transcriptomic changes during nematode-induced giant cell development in tomato compared with *Arabidopsis*: a functional role for gene repression. New Phytol 2013; 197(4): 1276-90.
[http://dx.doi.org/10.1111/nph.12121] [PMID: 23373862]

[100] Gommers FJ, Voorin'tholt DJM. Chemotaxonomy of Compositae related to their host suitability for *Pratylenchus penetrans.* Neth J Plant Pathol 1976; 82(1): 1-8.
[http://dx.doi.org/10.1007/BF01977341]

[101] Ali MA, Azeem F, Li H, Bohlmann H. Smart parasitic nematodes use multifaceted strategies to parasitize plants. Front Plant Sci 2017; 8: 1699.
[http://dx.doi.org/10.3389/fpls.2017.01699] [PMID: 29046680]

[102] Gheysen G, Mitchum MG. How nematodes manipulate plant development pathways for infection. Curr Opin Plant Biol 2011; 14(4): 415-21.
[http://dx.doi.org/10.1016/j.pbi.2011.03.012] [PMID: 21458361]

[103] Hewezi T. Cellular signaling pathways and posttranslational modifications mediated by nematode effector proteins. Plant Physiol 2015; 169(2): 1018-26.
[http://dx.doi.org/10.1104/pp.15.00923] [PMID: 26315856]

[104] Manosalva P, Manohar M, von Reuss SH, *et al.* Conserved nematode signalling molecules elicit plant defenses and pathogen resistance. Nat Commun 2015; 6(1): 7795.
[http://dx.doi.org/10.1038/ncomms8795] [PMID: 26203561]

[105] Lin B, Zhuo K, Chen S, *et al.* A novel nematode effector suppresses plant immunity by activating host reactive oxygen species-scavenging system. New Phytol 2016; 209(3): 1159-73.

[http://dx.doi.org/10.1111/nph.13701] [PMID: 26484653]

[106] Naalden D, Haegeman A, de Almeida-Engler J, Birhane Eshetu F, Bauters L, Gheysen G. The *Meloidogyne graminicola* effector Mg16820 is secreted in the apoplast and cytoplasm to suppress plant host defense responses. Mol Plant Pathol 2018; 19(11): 2416-30.
[http://dx.doi.org/10.1111/mpp.12719] [PMID: 30011122]

[107] Yang S, Dai Y, Chen Y, *et al.* A novel G16B09-like effector from *Heterodera avenae* suppresses plant defenses and promotes parasitism. Front Plant Sci 2019; 10: 66.
[http://dx.doi.org/10.3389/fpls.2019.00066] [PMID: 30800135]

[108] Mendy B, Wang'ombe MW, Radakovic ZS, *et al. Arabidopsis* leucine-rich repeat receptor–like kinase NILR1 is required for induction of innate immunity to parasitic nematodes. PLoS Pathog 2017; 13(4): e1006284.
[http://dx.doi.org/10.1371/journal.ppat.1006284] [PMID: 28406987]

[109] Haegeman A, Mantelin S, Jones JT, Gheysen G. Functional roles of effectors of plant-parasitic nematodes. Gene 2012; 492(1): 19-31.
[http://dx.doi.org/10.1016/j.gene.2011.10.040] [PMID: 22062000]

[110] Juvale PS, Baum TJ. "Cyst-ained" research into *Heterodera* parasitism. PLoS Pathog 2018; 14(2): e1006791.
[http://dx.doi.org/10.1371/journal.ppat.1006791] [PMID: 29389955]

[111] El-Sappah AH, M M I, H El-Awady H, *et al.* Tomato natural resistance genes in controlling the root-knot nematode. Genes (Basel) 2019; 10(11): 925.
[http://dx.doi.org/10.3390/genes10110925] [PMID: 31739481]

[112] Sato K, Kadota Y, Shirasu K. Plant immune responses to parasitic nematodes. Front Plant Sci 2019; 10: 1165.
[http://dx.doi.org/10.3389/fpls.2019.01165] [PMID: 31616453]

[113] Gao B, Allen R, Maier T, Davis EL, Baum TJ, Hussey RS. Identification of a new beta-1,4-endoglucanase gene expressed in the esophageal subventral gland cells of *Heterodera glycines.* J Nematol 2002; 34(1): 12-5.
[PMID: 19265901]

[114] Qin L, Kudla U, Roze EHA, *et al.* A nematode expansin acting on plants. Nature 2004; 427(6969): 30.
[http://dx.doi.org/10.1038/427030a] [PMID: 14702076]

[115] Kudla U, Qin L, Milac A, *et al.* Origin, distribution and 3D-modeling of Gr-EXPB1, an expansin from the potato cyst nematode *Globodera rostochiensis.* FEBS Lett 2005; 579(11): 2451-7.
[http://dx.doi.org/10.1016/j.febslet.2005.03.047] [PMID: 15848187]

[116] Opperman CH, Bird DM, Williamson VM, *et al.* Sequence and genetic map of *Meloidogyne hapla* : A compact nematode genome for plant parasitism. Proc Natl Acad Sci USA 2008; 105(39): 14802-7.
[http://dx.doi.org/10.1073/pnas.0805946105] [PMID: 18809916]

[117] Lozano-Torres JL, Wilbers RHP, Gawronski P, *et al.* Dual disease resistance mediated by the immune receptor Cf-2 in tomato requires a common virulence target of a fungus and a nematode. Proc Natl Acad Sci USA 2012; 109(25): 10119-24.
[http://dx.doi.org/10.1073/pnas.1202867109] [PMID: 22675118]

[118] Chen S, Wan Z, Nelson MN, *et al.* Evidence from genome-wide simple sequence repeat markers for a polyphyletic origin and secondary centers of genetic diversity of *Brassica juncea* in China and India. J Hered 2013; 104(3): 416-27.
[http://dx.doi.org/10.1093/jhered/est015] [PMID: 23519868]

[119] Kud J, Wang W, Gross R, *et al.* The potato cyst nematode effector RHA1B is a ubiquitin ligase and uses two distinct mechanisms to suppress plant immune signaling. PLoS Pathog 2019; 15(4): e1007720.
[http://dx.doi.org/10.1371/journal.ppat.1007720] [PMID: 30978251]

[120] Vieira P, Gleason C. Plant-parasitic nematode effectors — insights into their diversity and new tools for their identification. Curr Opin Plant Biol 2019; 50: 37-43.
[http://dx.doi.org/10.1016/j.pbi.2019.02.007] [PMID: 30921686]

[121] Jaouannet M, Perfus-Barbeoch L, Deleury E, *et al.* A root-knot nematode-secreted protein is injected into giant cells and targeted to the nuclei. New Phytol 2012; 194(4): 924-31.
[http://dx.doi.org/10.1111/j.1469-8137.2012.04164.x] [PMID: 22540860]

[122] Desmedt W, Mangelinckx S, Kyndt T, Vanholme B. A phytochemical perspective on plant defense against nematodes. Front Plant Sci 2020; 11: 602079.
[http://dx.doi.org/10.3389/fpls.2020.602079] [PMID: 33281858]

[122] Hewezi T, Piya S, Richard G, Rice JH. Spatial and temporal expression patterns of auxin response transcription factors in the syncytium induced by the beet cyst nematode *Heterodera schachtii* in *Arabidopsis*. Mol Plant Pathol 2014; 15(7): 730-6.
[http://dx.doi.org/10.1111/mpp.12121] [PMID: 24433277]

[123] Cabrera J, Fenoll C, Escobar C. Genes co-regulated with *LBD16* in nematode feeding sites inferred from *in silico* analysis show similarities to regulatory circuits mediated by the auxin/cytokinin balance in *Arabidopsis*. Plant Signal Behav 2015; 10(3): e990825.
[http://dx.doi.org/10.4161/15592324.2014.990825] [PMID: 25664644]

[124] De Meutter J, Tytgat T, Prinsen E, Gheysen G, Van Onckelen H, Gheysen G. Production of auxin and related compounds by the plant parasitic nematodes *Heterodera schachtii* and *Meloidogyne incognita*. Commun Agric Appl Biol Sci 2005; 70(1): 51-60.
[PMID: 16363359]

[125] Kyndt T, Goverse A, Haegeman A, *et al.* Redirection of auxin flow in *Arabidopsis thaliana* roots after infection by root-knot nematodes. J Exp Bot 2016; 67(15): 4559-70.
[http://dx.doi.org/10.1093/jxb/erw230] [PMID: 27312670]

[126] Lee C, Chronis D, Kenning C, *et al.* The novel cyst nematode effector protein 19C07 interacts with the *Arabidopsis* auxin influx transporter LAX3 to control feeding site development. Plant Physiol 2011; 155(2): 866-80.
[http://dx.doi.org/10.1104/pp.110.167197] [PMID: 21156858]

[127] De Meutter J, Tytgat T, Witters E, Gheysen G, Van Onckelen H, Gheysen G. Identification of cytokinins produced by the plant parasitic nematodes *Heterodera schachtii* and *Meloidogyne incognita*. Mol Plant Pathol 2003; 4(4): 271-7.
[http://dx.doi.org/10.1046/j.1364-3703.2003.00176.x] [PMID: 20569387]

[128] Siddique S, Radakovic ZS, De La Torre CM, *et al.* A parasitic nematode releases cytokinin that controls cell division and orchestrates feeding site formation in host plants. Proc Natl Acad Sci USA 2015; 112(41): 12669-74.
[http://dx.doi.org/10.1073/pnas.1503657112] [PMID: 26417108]

[129] Fudali SL, Wang C, Williamson VM. Ethylene signaling pathway modulates attractiveness of host roots to the root-knot nematode *Meloidogyne hapla*. Mol Plant Microbe Interact 2013; 26(1): 75-86.
[http://dx.doi.org/10.1094/MPMI-05-12-0107-R] [PMID: 22712507]

[130] Kammerhofer N, Radakovic Z, Regis JMA, *et al.* Role of stress-related hormones in plant defence during early infection of the cyst nematode *Heterodera schachtii* in *Arabidopsis*. New Phytol 2015; 207(3): 778-89.
[http://dx.doi.org/10.1111/nph.13395] [PMID: 25825039]

[132] Piya S, Binder BM, Hewezi T. Canonical and noncanonical ethylene signaling pathways that regulate *Arabidopsis* susceptibility to the cyst nematode *Heterodera schachtii*. New Phytol 2018.
[http://dx.doi.org/10.1111/nph.15400] [PMID: 30136723]

[133] Gheysen G, Mitchum MG. Phytoparasitic nematode control of plant hormone pathways. Plant Physiol 2019; 179(4): 1212-26.

[http://dx.doi.org/10.1104/pp.18.01067] [PMID: 30397024]

[134] Gheysen G, Mitchum MG. Phytoparasitic nematode control of plant hormone pathways. Plant Physiology 2019; 179(4): 1212-26.
[http://dx.doi.org/10.1104/pp.18.01067]

CHAPTER 6

Environmental Factors Influencing Phytochemical Production for Enhanced Phytochemical Defense

Ishfaq Majeed Malik[1], Aashaq Hussain Bhat[3,*], Danish Majeed[2] and **Naveed Nabi[3]**

[1] *Department of Zoology, University of Kashmir, Hazratbal 190006, Srinagar, Jammu and Kashmir, India*

[2] *Department of Botany, University of Kashmir, Hazratbal 190006, Srinagar, Jammu and Kashmir, India*

[3] *Department of Biosciences, University Centre for Research and Development, Chandigarh University, Gharuan, Mohali 140413, Punjab, India*

Abstract: Phytochemicals are essential compounds in plants that serve as advanced defense mechanisms against various environmental stressors. This chapter delves into the environmental factors influencing phytochemical biosynthesis, providing a thorough analysis of how plants adapt to different stress conditions. Both abiotic and biotic stressors have a significant impact on phytochemical production. Abiotic stressors, such as temperature fluctuations, variations in light intensity and spectrum, water availability, soil conditions, and salinity, can distinctly modify phytochemical profiles. Extreme temperatures can alter the composition of phytochemicals, while light conditions, including photoperiod and wavelength, regulate the synthesis of crucial compounds. Water stress, from drought or waterlogging, affects phytochemical compositions, and soil factors like pH and nutrient levels influence the overall phytochemical profile. Saline environments induce osmotic stress, leading to notable changes in phytochemical production. Biotic stressors, including pathogen attacks, herbivory, and competitive interactions, also significantly impact phytochemical synthesis. Plants generate induced defenses in response to pathogens, and secondary metabolites play a crucial role in deterring herbivores. Competitive interactions, such as allelopathy, influence phytochemical production, highlighting the complexity of plant responses in competitive settings. The chapter also explores methods to enhance phytochemical production through environmental modulation. Agricultural practices like crop rotation, intercropping, and organic farming can boost phytochemical content. Controlled environment agriculture, such as greenhouse and hydroponic systems, optimizes conditions for superior phytochemical synthesis. Additionally, genetic and biotechnological advancements, including genetic engineering, plant breeding, and the

---

* **Corresponding author Aashaq Hussain Bhat:** Department of Biosciences, University Centre for Research and Development, Chandigarh University, Gharuan, Mohali 140413, Punjab, India;
E-mail: aashiqhussainbhat10@gmail.com

use of elicitors and biostimulants, offer promising avenues for increasing phytochemical yields. Future research should focus on refining agricultural practices, optimizing controlled environments, and leveraging genetic and biotechnological innovations to enhance phytochemical production, promoting sustainable agriculture and strengthening plant resilience.

Keywords: Abiotic stress, Biotechnological advancements, Biotic stress, Controlled environment agriculture, Phytochemicals, Phytochemical biosynthesis, Plant resilience.

INTRODUCTION

Plants constantly face a dynamic and often hostile environment, contending with numerous biotic stresses such as attacks from bacteria, fungi, viruses, insects, and herbivores, as well as abiotic stresses including UV radiation, drought, and soil contaminants [1]. In response, they have evolved a sophisticated array of phytochemicals that act as their natural defense mechanisms. Plants, despite being stationary, have evolved to produce a vast array of secondary metabolites. These organic compounds, while not directly involved in primary growth and development, play crucial roles in the survival and resilience of plants [2]. These phytochemicals serve various ecological functions, primarily in plant defense, and encompass a diverse range of classes such as alkaloids, phenolics, terpenoids, and flavonoids, each uniquely contributing to plant defense strategies. Phytochemicals form the first line of defense against pathogen invasion and herbivore feeding through multiple mechanisms [3]. Some possess direct antimicrobial properties, while others work indirectly by attracting beneficial organisms or inducing systemic resistance within the plant [4]. Additionally, these compounds help plants cope with environmental stresses by mitigating UV radiation damage, detoxifying harmful substances, and enhancing nutrient uptake. Understanding the multifaceted roles of phytochemicals in plant defense not only illuminates the complex ways plants interact with their environment but also paves the way for agricultural innovation. Harnessing the natural defensive properties of phytochemicals can foster the development of more resilient crops, thereby reducing reliance on synthetic pesticides and promoting sustainable agricultural practices. Thus, the study of phytochemicals is crucial for ensuring food security and environmental sustainability in the face of global challenges.

Secondary metabolites, which are mainly classified into phenolics, terpenes, and nitrogen/sulfur-containing compounds, are essential for plant defense against predators and pathogens. Terpenes, constructed from 5-carbon isoprene units, function as toxins to deter herbivores. Phenolics, synthesized *via* the shikimic acid pathway, bolster the plant's defense mechanisms. Additionally, nitrogen and sulfur compounds, which derive from amino acids, contribute significantly to plant protection. *In vitro* studies have confirmed the defensive roles of these metabolites, revealing over 100,000 known compounds involved in plant defense, although the full extent of their diversity remains unclear [5]. Higher concentrations of secondary metabolites typically enhance plant resistance to both biotic and abiotic stresses. However, their production can impose significant costs on plant growth and reproduction [5]. The evolution of induced defense mechanisms, characterized by increased concentrations of secondary metabolites under stress conditions, highlights their structural and functional significance. Numerous studies have identified a wide array of plant compounds with ecological and chemical defensive roles, forming the basis of the emerging field of ecological biochemistry. Exploring phytochemicals in plant defense not only deepens our understanding of plant-environment interactions but also offers valuable insights for developing sustainable agricultural practices and resilient crop varieties. Furthermore, primary metabolites, such as proteins, fats, carbohydrates, and dietary fiber, are essential for energy metabolism and cell structure. In contrast, secondary metabolites, which are non-nutritive, play a vital role in plant-environment interactions, including defense against insects and fungi (Fig. **1**).

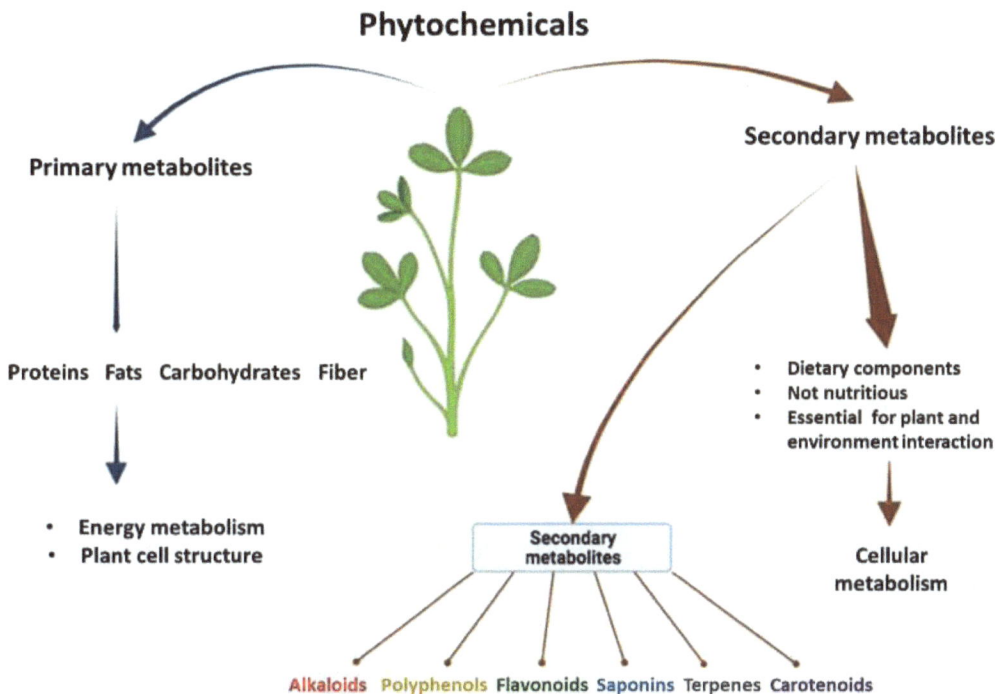

Fig. (1). The difference between primary and secondary metabolites.
(Source: https://doi.org/10.3390/plants13040523).

IMPACT OF PLANT SECONDARY METABOLITES ON DEFENSE AGAINST PATHOGENS

Secondary metabolites have been widely studied for their interaction with pathogenic organisms, making them a critical component of plant immune responses [6, 7]. Over decades of research, it has been established that these metabolites play significant roles in defending plants against pathogens. Despite their structural diversity, these compounds operate within a conserved framework, even though several biosynthetic pathways vary across plant species [8]. The production and activation of secondary metabolites are initiated when plants detect microbes. This detection occurs through defense proteins or the recognition of microbe-associated molecular patterns (MAMPs) by pattern recognition receptors (PRRs). Among the wide array of secondary metabolites, phenylpropanoids, and flavonoids are prominent due to their diverse modes of action. These compounds interfere with critical processes within pathogens, such as cell signaling, physiological activities, and key components like enzymes, DNA, and reproductive systems [9]. The phenolic hydroxyl groups in these compounds often dissociate into phenolate ions, which form ionic and hydrogen bonds with peptides and protons. This leads to high astringency and protein denaturation when present in large numbers [10]. Proteins require specific three-dimensional structures to function correctly; any alteration in their conformation can change their properties and inhibit interactions with other proteins and nucleic acids. Secondary metabolites can modify protein structures by forming covalent bonds with free SH, OH⁻, or amino groups, leading to functional loss or altered protein turnover. Polyphenols, in particular, form hydrogen bonds and strong ionic bonds that can inactivate proteins by altering their flexibility through weak non-covalent interactions. Their polar nature makes phenols less toxic, as their absorption after oral intake is low numbers.

Phenolic hydroxyl groups in these compounds often dissociate into phenolate ions. These ions can form ionic and hydrogen bonds with peptides and protons, leading to astringency and protein denaturation when present in large quantities [10]. Proteins require specific three-dimensional structures to function correctly; any change in their conformation can alter their properties and inhibit interactions with other proteins and nucleic acids. Secondary metabolites can modify protein structures by forming covalent bonds with free SH, OH-, or amino groups, resulting in functional loss or altered protein turnover. Polyphenols, in particular, can form hydrogen and strong ionic bonds that inactivate proteins by altering their flexibility through weak non-covalent interactions. Their polar nature makes phenols less toxic, as their absorption after oral intake is minimal [10].

Plants have developed a sophisticated system to recognize and respond to herbivore attacks, ensuring early detection and activation of robust defense mechanisms [11]. Recent genetic and chemical research has highlighted the multifunctional roles of PSMs, which are crucial for regulating plant growth, defense, and primary metabolism. Induced plant defenses are primarily driven by the phytohormones jasmonic acid (JA) and salicylic acid (SA), which interact in complex ways at both transcript and protein levels. Studies have shown that applications of JA and SA can negatively affect the survival of both chewing insects (*Heliothis virescens*) and sucking insects (*Myzus persicae*) [12]. PSMs may also function similarly to hormones by binding to specific receptor proteins, working synergistically to counteract herbivore damage and potentially delaying resistance development in insect herbivores [13]. However, insects can adapt through mechanisms such as detoxification, excretion, or sequestration of PSMs [14]. While the defensive role of secondary metabolites is well-documented, some also attract insect pollinators and parasitoids, enhancing ecological interactions. These metabolites are considered a cost-effective and environmentally friendly alternative to agrochemicals in agriculture, as they significantly contribute to plant growth and protection [15]. Thus, secondary metabolites are essential for plant defense, offering diverse mechanisms to protect against pathogens and herbivores while also promoting ecological interactions and agricultural sustainability.

INFLUENCE OF ENVIRONMENTAL CONDITIONS ON PHYTOCHEMICAL SYNTHESIS

Environmental factors play a crucial role in the production of PSMs, which are essential for the color, taste, odor, and defensive responses of plants [16]. The production and concentration of these metabolites vary significantly both among and within plant species due to multiple influences [17]. Various cellular and biochemical factors impact the storage and transport of PSMs, while developmental factors influence their biosynthesis and storage [18]. Environmental stresses—such as nutrient deficiencies, physical damage, metal ions, UV radiation, light, circadian rhythms, seasonal changes, salinity, drought, and temperature—play a significant role in modulating PSM levels [16, 19]. Furthermore, the metabolic pathways and growth conditions of plants influence PSM concentrations [20]. Biotic factors, such as pathogen attacks, also trigger plant defense mechanisms, resulting in variations in PSM levels in response to environmental stresses like light intensity and nutrient availability [16]. Genetic factors contribute by affecting gene expression related to PSM biosynthesis. *In vitro*, tissue culture techniques can enhance PSM production using signaling molecules and plant growth regulators. Overall, environmental, morphogenetic, ontogenetic, and genetic factors influence PSM levels, with environmental factors being the most significant [16, 21].

Plants exhibit variability in the biosynthesis and accumulation of secondary metabolites (SMs) in response to abiotic and biotic factors, which significantly influence SM production [22]. SM concentrations can differ among plants of the same species when exposed to varying environmental conditions [23]. Environmental stresses stimulate SM production, aiding plants in mitigating adverse effects. Abiotic stresses, such as soil composition, temperature extremes, light intensity, and drought, significantly impact plant growth and productivity, while biotic stresses caused by living organisms also reduce plant growth and yield [23]. Abiotic stress encompasses a broad range of factors, including radiation, nutrient deficiencies, pesticides, heavy metals, pollutants, toxic gases like ozone, and salinity.

Abiotic stresses include factors like chemical fertilizers, soil type and composition, temperature fluctuations, light intensity, water availability, and salinity, affecting plants throughout their developmental stages. Optimal levels of these abiotic factors are essential for plant growth; any excess or deficiency can alter the synthesis of SMs, crucial for plant productivity [16]. Phenylpropanoids, a significant class of SMs, often increase in response to environmental stresses like temperature extremes, nutrient deficiencies, wounding, and UV radiation [24]. Understanding the interplay of these factors is crucial for optimizing PSM production for various applications.

Biotic stresses arise from living organisms such as nematodes, bacteria, viruses, and fungi. Plants, being immobile, display high tolerance to pathogen attacks primarily through SM production. Phytoalexins, a type of SM, enhance plant defense responses due to their antimicrobial properties [25]. Pathogen attacks boost SM biosynthesis; for instance, *Lupinus angustifolius* infected by fungal strains show significant variations in endogenous phenolic levels [16]. Pathogen attacks induce an innate immune system in plants mediated by two mechanisms: effector-triggered immunity and basal immunity. Basal immunity involves the detection of pathogens through microbe-associated molecular patterns recognized by pattern recognition receptors on host cells, while effector-triggered immunity responds to pathogen effectors or toxins. Plants perceive these signals and activate various metabolic pathways to produce SMs, with their concentration decreasing significantly during stress recovery [26].

IMPACT OF ABIOTIC STRESSORS ON PHYTOCHEMICAL PRODUCTION

Influence of Temperature on Phytochemical Production

The impact of rising temperatures is a significant concern, as highlighted in projections by Masson-Delmotte *et al.*, 2021 [27]. High temperatures (HT) are known to diminish crop yields by lowering net photosynthesis, increasing transpiration, and elevating stomatal conductance [28]. HT can visibly harm crops, causing scorching of leaves and stems, leaf shedding, and premature aging [29]. Specific temperature ranges are crucial for processes such as seed germination and flowering, and extreme temperature events have historically resulted in substantial economic losses, posing a threat to global food security [30]. For instance, HT is projected to reduce potato yields by 18-32%, making it the third most important global food crop [31]. Additionally, short-term heat stress can permanently affect the biochemistry of developing anthers in barley and *Arabidopsis thaliana*, leading to male sterility [32].

Unlike other stresses, temperature affects all cellular components simultaneously through thermosensors that initiate downstream biochemical responses [32]. While the precise mechanisms remain unclear, heat shock transcription factor A1s (HsfA1s) play a pivotal role in the plant heat stress response [30]. Plants also face cold stress at temperatures below 20°C and freezing stress at sub-zero temperatures, both of which negatively impact plant growth and distribution, causing agricultural losses [33]. Temperature fluctuations trigger cellular responses that significantly alter plant metabolism [31, 34, 35]. These stresses can lead to protein denaturation, lipid destabilization, and disruption of membrane integrity [35]. In response, plants reduce energy and protein metabolism to accumulate protective compounds [31], indicating a temperature-dependent regulation of PSMs [36]. Thermal responses are rapid, with most temperature-related changes in metabolites occurring within the first 30 minutes of exposure [33]. The shift of carbon from growth to defense [35] may indicate maladaptive effects; while plants may survive, they often become smaller and yield less. Increasing evidence suggests that damage to cellular membranes at extreme temperatures is linked to increased production of highly toxic reactive oxygen species (ROS), with lipid peroxidation serving as an indicator of oxidative stress [37]. Thermotolerance involves the synergistic effects of various compatible solutes, including hormones and several PSMs. While both heat and cold stress trigger thermal signaling, these responses may follow different pathways [33]. In *Arabidopsis*, among 497 measured metabolites, 31% showed no response to either temperature extreme, 4% responded only to heat, 19% to both extremes and 38% only to cold [33]. Therefore, it is crucial to acknowledge that a PSM that increases in response to HT may not necessarily decrease in response to cold stress, necessitating separate analyses of these extremes. Understanding the nuanced effects of temperature extremes on plant secondary metabolite production is essential for developing strategies to mitigate the adverse impacts of climate change on agriculture.

Influence of Low-Temperature Stress on Phytochemicals Production

Cold stress significantly impacts the production of PSMs, often directing shikimic acid toward the shikimate pathway, thereby increasing metabolite pools. For instance, kale shows increased levels of phenolic compounds like flavonol, quercetin, kaempferol, and isorhamnetin under cold stress [38]. Similarly, grapevine exhibits the highest flavonol content at lower temperatures [39], while apple sees rises in anthocyanins and flavonoids [40]. This enhancement in flavonoid biosynthesis may be triggered by elevated sucrose levels acting as a signaling molecule in colder conditions [41]. Conversely, Siberian ginseng shows decreased total phenolic content at low temperatures [37], as do pygmy smartweed in flavanol content [42] and St. John's wort in hypericin levels [35], indicating species-specific responses.

Most terpenoids, contributing to the bitter taste in crops like carrots, generally decrease at low temperatures (9°C or 12°C), except for β-terpinolene, which increases [43]. Asters show an increase in the sesquiterpene artemisinin under cold stress [44], potentially impacting plant palatability. Regarding nitrogen-containing PSMs, cold stress enhances quercetin glycosides but reduces kaempferol glycosides in kale, and increases alkyl glucosinolates such as glucoiberin and glucoraphanin in broccoli. Additionally, cold acclimation raises levels of the non-protein amino acid GABA in castor beans [34] and *Arabidopsis* [33], likely due to increased glutamate decarboxylase activity in the GABA biosynthetic pathway [34]. Thus, cold stress exerts diverse and species-specific effects on PSMs, influencing various metabolic pathways and resulting in significant changes in phenolics, flavonoids, terpenoids, and nitrogen-containing compounds. Understanding these responses is crucial for optimizing crop quality and enhancing stress resilience.

Impact of Thermal Stress on Phytochemical Production

Heat stress, similar to cold stress, induces increased accumulation of polyamines in plants. In heat-sensitive rice cultivars, elevated temperatures lead to higher levels of putrescine, which subsequently enhance levels of spermidine and spermine. These polyamines aid in scavenging radicals and may improve metabolic flux through the Krebs cycle in temperature-sensitive cultivars [45]. In castor beans, higher temperatures decrease shikimate concentrations, thereby promoting the formation of downstream amino acid derivatives such as tryptophan, tyrosine, and phenylalanine [34]. Siberian ginseng grown at 24°C exhibits the highest levels of total phenolics and flavonoids, but these levels significantly decrease at 30°C [37]. Similarly, sugarcane and grapevines show reduced flavonoid and anthocyanin contents, respectively, under heat stress [29, 39].

Mung beans exposed to higher temperatures exhibit lower flavonoid concentrations, likely due to disruption in flavonoid biosynthesis under extreme heat conditions [46]. European aspen saplings also display varied responses to temperature; male genotypes reduce phenolic biosynthesis, while female genotypes show a dilution effect [47]. Overall, phenolic compounds play crucial roles as adaptive reactive oxygen species (ROS) scavengers during heat stress, responding differently across biosynthetic pathways. Heat stress can also upregulate key enzymes, increasing isoprene production in plants. For instance, carrots grown at 21°C emit more minor terpenes, while major terpenes remain stable at lower temperatures [43]. Sugarcane and rice show enhanced carotenoid and terpene levels under heat stress, respectively, while broccoli exhibits decreased lutein and β-carotene at temperatures above 15°C [36]. Understanding these metabolic responses across plant species can aid in developing strategies to enhance plant resilience to temperature fluctuations.

Impact of Drought Stress on Phytochemical Production

Drought stress (DS) results from reduced water availability due to low precipitation or increased evaporation [48], causing decreased plant growth and yield [49]. Agriculture, consuming over 70% of available water with more than 40% relying on irrigation, faces substantial economic impacts from changes in water supply [50]. While severe DS can harm plants, some species can acclimate over weeks, undergoing metabolic adjustments that lead to diminished growth [49, 51].

Plant responses to DS are species-specific, involving the accumulation of solutes like proline and polyols to counteract osmotic stress and protect against protein degradation [52]. Flavonoids typically increase in response to DS in about half of the cases, while non-flavonoid phenolic compounds rise in two-thirds, primarily in drought-tolerant cultivars. For example, in maize, only tolerant genotypes showed an increase in total phenolic content [53]. Terpenoid responses to drought vary; about 70% of terpenes do not increase under DS, with peak levels observed approximately two weeks post-stress in some species like thyme, though declining later [51].

In liquorice, glycyrrhizin and genes related to triterpene saponins increase under DS, contrasting with the rise of monoterpenes in thyme. Terpene emissions decrease due to DS's impact on glandular trichomes, which store these compounds. Carotenoids generally decline under stress, except in specific cases like *Nasturtium*. Soil water content critically regulates nitrogen-containing PSMs, influencing crop taste in plants such as mustard and horseradish [54]. For example, DS approaching the permanent wilting point in *Nasturtium* increases glucosinolate levels, impacting crops like canola and Ethiopian kale [54]. DS also elevates cyanogenic glycosides in lima beans, highlighting its significant implications for crop growth [55].

Water scarcity affects plants by reducing turgor pressure and water potential, disrupting physiological processes [56]. This stress suppresses photosynthesis and growth, leading to biochemical changes such as disrupted enzyme activities and membrane integrity loss, along with stomatal closure. Interestingly, several secondary metabolites (SMs) produced by plants contribute to drought tolerance [16]. Various medicinal plants, including *Catharanthus roseus*, *Hypericum perforatum*, and *Artemisia annua*, exhibit increased endogenous SM levels in response to drought stress. For instance, *Trachyspermum ammi* showed increased phenolic compounds and photosynthetic pigments but decreased biomass under drought [57].

In *Hypericum brasiliense*, drought stress enhanced the quality of crucial SMs such as rutin, quercetin, and betulinic acid, while *Artemisia* species showed increased artemisinin levels [16]. St. John's wort exposed to water-limited conditions exhibited reduced photosynthesis but increased concentrations of SMs like pseudohypericin, hypericin, and hyperforin [35]. Additionally, *Glechoma longituba* demonstrated increased total flavonoids under drought conditions [58]. Water scarcity in *Ocimum americanum* and *Ocimum basilicum* notably affected macronutrient levels, proline, carbohydrates, and essential oils [59]. Understanding plant responses to drought stress is crucial for developing strategies to enhance crop resilience and productivity under changing climatic conditions.

Impact of Light on Phytochemical Production

Light profoundly influences plant growth and success through factors such as intensity, duration, and quality [60, 61]. Plants possess photoreceptors that enable them to detect changes in light [62] and absorb approximately 90% of red and blue light, making them highly responsive to their light environment [63]. Both light quantity and quality significantly impact plant morphology, development, and the synthesis of primary and secondary metabolites in a species-specific manner. Light is essential for crucial cellular processes including hypocotyl expansion, chloroplast development, leaf growth, and flowering initiation [64]. Blue light, in particular, stimulates phototropism and regulates stem length, chloroplast movement, stomatal function, and gene expression. While light intensity generally enhances the accumulation of phytochemicals, light quality can evoke more nuanced, species-specific responses [65].

Key metabolites like phenolic compounds, activated by light, serve as antimicrobial agents and antioxidants that scavenge reactive oxygen species (ROS) [66]. Carotenoids, crucial pigments in chloroplast thylakoid membranes, optimize energy capture and dissipate excess energy as heat through interaction with photosystem complexes. Light-induced PSMs can act as direct defense compounds or precursors for biosynthetic pathways. For example, increased light exposure in *Arabidopsis* boosts the accumulation of monolignol glucosides, crucial for lignin biosynthesis [65]. Under light stress conditions, specific PSMs accumulate in targeted plant areas; for instance, antioxidant coumarins gather in outer vacuolar layers of sun-exposed leaves in manna ash, while less effective antioxidants are deeper within [67]. Overall, light plays a pivotal role in plant metabolism, enhancing carbon-to-nitrogen balance and boosting the production of secondary metabolites, even under conditions of limited photosynthetic capacity.

Impact of Light Quantity on Phytochemical Production

Light quantity significantly influences the biosynthesis of phytochemicals, operating within a diurnal cycle and highlighting the role of photoperiod alongside light intensity. Studies have shown varied effects on phenolic production across different plant species. For instance, while *Arabidopsis* and potato demonstrate increased levels of individual phenolics under heightened light, total phenolics decrease in species like *Arabidopsis*, cat's whiskers, and shade-intolerant jewel orchids [66, 68, 69]. This suggests that sensitive species may experience reduced phytochemical production under high light levels. Conversely, sweet basil and buckwheat exhibit enhanced phenolic accumulation with longer photoperiods, highlighting species-specific responses [70, 71].

The influence of light on other phytochemical groups such as terpenoids and nitrogen-containing compounds is also notable. Isoprene emissions, for example, increase during transient events like droughts or heat waves, which are often associated with changes in light availability. Carotenoid levels in field mustard increase under high light conditions but decrease in Indian mustard, reflecting their protective role in photosynthesis [72]. Similarly, the response of nitrogen-containing compounds varies; aliphatic glucosinolates in wild cabbage increase with longer photoperiods, while indole glucosinolates remain unaffected [73]. Plants like white mustard also show increased glucosinolate levels with extended photoperiods [74]. Moreover, extended photoperiods enhance the production of compounds such as GABA, glycosides, and alkaloids in various plants, including ginsenosides in American ginseng and solanine in potatoes [75, 76]. Understanding these light-mediated effects on secondary metabolite production is crucial for optimizing growth conditions and improving the medicinal and nutritional properties of plants.

Impact of Light Quality on Phytochemical Production

Research into the impact of light quality on PSMs reveals significant variability among species. This variability is influenced by the ratios of blue, red, and far-red light, as well as the type of light source used (*e.g.*, incandescent, fluorescent). For instance, blue light consistently increases flavonoid content and other phenolic compounds across various species, potentially preparing plants for environmental stresses [61]. On the other hand, red light triggers the production of specific metabolites such as rosmarinic acid in basil and p-coumaric acid in ginseng [71].

Studies also indicate that yellow and green lights can elevate phenolics and flavonoids in plants like self-heal, whereas green light enhances the production of terpenes such as perillaldehyde and limonene, often with a trade-off of reduced plant biomass [77]. Additionally, blue light promotes nitrogen-containing compounds like glucosinolates and glycosides, with notable increases observed in broccoli [62]. Red light, on the other hand, fosters the accumulation of specific glycosides in Siberian ginseng and gluconasturtiin in watercress [37, 77]. Interestingly, the response of biomass to blue light varies among species and cultivars, suggesting differential partitioning of assimilates due to varying enzymatic activities. Therefore, understanding how different light qualities modulate PSMs is crucial for optimizing plant growth and enhancing the phytochemical composition of agricultural and medicinal plants.

Impact of Radiation Stress on Phytochemical Production

Plants depend on optimal light conditions to facilitate photosynthesis, a vital process for the synthesis of secondary metabolites (SMs). Research on *Mikania glomerata* illustrates that sunlight plays a critical role in enhancing the production of coumarins. The duration of light exposure significantly affects coumarin levels across different plant parts [78]. Shorter light periods decrease coumarin content, whereas prolonged exposure leads to higher concentrations in leaves and stems. This highlights how photoperiod and light intensity profoundly influence the biosynthesis and accumulation of SMs [16]. Light serves as a crucial abiotic factor that directly influences plant growth, photosynthesis, and consequently, the quantity and quality of SMs. Understanding these factors is essential for optimizing the production of valuable plant compounds.

Impact of Salinity Stress on Phytochemical Production

Salinity stress arises from elevated salt levels in growth media, inducing osmotic stress in plants, which limits water absorption despite its availability [79]. This condition adversely affects photosynthesis, growth, and nutrient uptake. Plants respond to salinity stress by modulating the concentrations of secondary metabolites (SMs), which can either increase or decrease due to osmotic stress or specific ion toxicity [20].

For instance, *Catharanthus roseus* and *Rauvolfia tetraphylla* accumulate significant levels of vincristine alkaloids and reserpine under salinity stress [80]. Similarly, *Achillea fragrantissima*, *C. roseus*, and *Solanum nigrum* exhibit elevated alkaloid levels under salinity stress, with *A. fragrantissima* also showing increased endogenous phenolic levels [16]. *Matricaria chamomilla* responds by producing higher amounts of phenolics such as caffeic, chlorogenic, and protocatechuic acids under salinity stress [81]. Despite varying effects on essential oil production—decreasing in species like *Mentha suaveolens* and increasing in others like *Salvia officinalis*—salinity stress consistently enhances alkaloid levels, antioxidant potential, and the production of compounds like saponins and flavonoids across various plants [82, 83]. These adaptive responses underscore the intricate biochemical adjustments plants undergo to cope with salinity stress, emphasizing their significance in agricultural and ecological contexts.

RESPONSE OF PSMS TO SOIL FERTILITY

Recent studies have highlighted the profound influence of nutrient availability on plant secondary metabolism and antioxidant activity. Most of these investigations utilized highly water-soluble fertilizers, primarily rich in nitrogen (N), phosphorus (P), or potassium (K) [84]. Nutrient deficiencies, particularly nitrogen and phosphate, have been linked to increased flavonoid accumulation, notably anthocyanins, in plants such as *Arabidopsis* and tomato seedlings [85]. Under nitrogen or phosphate stress, species like these show elevated levels of quercetin, kaempferol, and isorhamnetin. Moreover, nitrogen levels significantly impact total phenolics and flavonoids in *Labisia pumila* Benth [86]. Furthermore, nutrient deficiencies in soil affect alkaloid concentrations, exemplified by *Lupinus angustifolius* seeds. Severe potassium deficiency can lead to a dramatic 205% increase in alkaloid concentrations in sweet lupin varieties, with variations observed across different alkaloids [87].

Studies on Douglas fir (*Pseudotsuga menziesii*) and *Thuja plicata* have shown that higher concentrations of ammonium nitrate and nitrogen fertilizers can stimulate monoterpene production, highlighting seasonal and species-specific responses [88, 89]. These findings underscore the complex relationship between soil nutrient availability and plant secondary metabolite synthesis. Thus, understanding how soil fertility influences PSMs is crucial for optimizing agricultural practices and enhancing plant resilience and productivity.

IMPACT OF BIOTIC STRESSORS ON PHYTOCHEMICAL PRODUCTION

A wide range of environmental stressors, both biotic and abiotic, influence plant metabolism ultimately affecting the agricultural crop output. Droughts, salinity, floods, heavy metals, severe temperatures, and other physical or chemical variables can contribute to abiotic stress [90]. In contrast, living creatures including bacteria, fungus, oomycetes, herbivores, insects, arachnids, nematodes, and weeds cause biotic stress [91]. Many parasitic fungi can either kill plant cells by releasing toxins (necrotrophic) or sustain themselves by feeding on live cells (biotrophic), which results in ailments like vascular wilts, cankers, and leaf spots in affected plants [92, 93]. On the other hand, nematodes, which feed on plant tissues, mostly cause soil-borne diseases that stunt growth, cause nutrient deficiencies, and lead to wilting [94]. Additionally, viruses are responsible for symptoms such as chlorosis and stunted growth through both localized and systemic damage [95]. Besides insects including mites threaten plants by sucking fluids, piercing tissues, and laying eggs, while also acting as carriers for pathogenic elements such as viruses and bacteria. To counter these stresses, plants have evolved with sophisticated defense systems [96]. They possess an initial, passive defense mechanism that includes physical barriers like cuticles, wax, and trichomes to deter insects and other pathogenic elements. In addition to the physical barriers, plants also synthesize chemical compounds as a defense strategy against invading pathogens [25]. Secondary metabolites are chemical compounds synthesized by plants to augment their defensive capability without any direct effect on their growth and development. They decrease the palatability of the tissues in which they are synthesized [97]. These chemical agents are either stored constitutively in inactive forms or their synthesis is induced in response to attacks by insects or microbes. The inactively stored chemical entities constitute phytoanticipins, which are typically activated by enzymes like β-glucosidases during herbivory, releasing biocidal aglycone metabolites [98]. Phytoanticipins include glucosinolates, which are hydrolyzed by myrosinases in the damaged tissue, and Benzoxazinoids (BXs), found widely in Poaceae, whose glucosides are hydrolyzed to produce defensive aglycone BXs [98]. Phytoalexins, on the other hand, encompass compounds like isoflavonoids, terpenoids, and alkaloids that affect the performance and survival of herbivores [99]. These secondary metabolites not only protect plants from various stresses but also enhance their overall fitness. For instance, in maize, resistance to the corn earworm, *Helicoverpa zea*, is largely attributed to secondary metabolites such as the Cglycosyl flavone maysin and the phenylpropanoid chlorogenic acid. Therefore, plants have developed intricate sensory systems to detect biotic intrusions and mitigate the impact on their growth, productivity, and survival [92, 100].

PHYTOCHEMICAL RESPONSE AGAINST PATHOGEN ATTACK: MECHANISMS INVOLVED

Many plant species produce diverse toxic secondary metabolites to combat pathogens. Secondary metabolites serve multiple functions in plant defense, acting as deterrents, toxins, and precursors to physical defense mechanisms. Host plants undergo metabolic changes that involve synthesizing antimicrobial proteins, enzymes, and various metabolites [101]. These responses bolster the plant's structural integrity and reduce damage inflicted by pathogens. Plant defense mechanisms against herbivores rely on intricate signal transduction pathways governed by a network of phytohormones. These hormones not only regulate plant growth and development but also orchestrate defense responses against herbivory. Several plant hormones like jasmonic acid (JA), salicylic acid (SA), and ethylene (ET) are known to participate in both intra- and inter-plant communication, particularly in plants that have been damaged by herbivores [99]. Upon herbivore-induced wounding or feeding, these signal-transduction pathways trigger specific sets of defense-related genes [102]. Depending on the type of attacker and the specific circumstances, these hormones may act individually, synergistically, or antagonistically to orchestrate the plant's response. The following section explores key plant metabolites crucial to these defense mechanisms.

Role of Melatonin

Melatonin exhibits a notable presence across all biological kingdoms, highlighting its potential to facilitate interactions between different kingdoms. Studies have shown that melatonin treatment enhances plant resistance to viral, bacterial, and fungal pathogens, particularly at low application levels [103, 104]. Mechanistically, melatonin contributes to plant defenses by perceiving threats, activating signaling and phytohormone pathways, stimulating innate immune responses, and bolstering the production of defensive compounds. Melatonin upregulates transcription factors involved in plant defense, triggers innate immunity responses, and induces the overproduction of sugars and glycerol associated with resistance mechanisms in plants like *Arabidopsis* [105, 106]. Conversely, pathogens can inhibit host melatonin production, potentially enhancing plant susceptibility to infections [107]. Moreover, it has been found that grapevine root bacterial symbionts like *Bacillus amyloliquefaciens* SB-9 secrete melatonin and induce its biosynthesis in plants, enhancing stress tolerance

[108]. Similarly, arbuscular mycorrhizal symbioses in legumes, such as alfalfa, show increased melatonin levels under stress [109]. Melatonin's role extends to mediating responses to wounding, insect feeding, and interactions between pathogens and plant resistance mechanisms [110]. For instance, in American elm populations affected by Dutch elm disease, melatonin signaling interacts with jasmonic acid and salicylic acid pathways, influencing resistance dynamics [111]. Melatonin application reduced the population of *Pseudomonas syringae*, a virulent bacterial pathogen, in infected leaves of *Arabidopsis thaliana*. Further research has highlighted the roles of melatonin, particularly serotonin [112] and N-acetyl serotonin [113], in elicitors that induce the expression of various defense-related genes against *P. syringae* in *Arabidopsis* and tobacco [114]. These insights were further confirmed using various *Arabidopsis* mutants, underscoring the role of melatonin in initiating the signaling pathways of defense genes upstream, which leads to the synthesis of phytohormones such as salicylic acid, jasmonic acid, and ethylene [115]. Altogether, these processes coordinate positively to enhance disease resistance in plants.

Protective Role of Flavonoids

Flavonoids play crucial roles in plant defense against pathogens and contribute to the chiefly to flower pigmentation [116]. They serve as secondary systems for scavenging of reactive oxygen species (ROS), thereby protecting photosynthetic pigments from damage due to excess excitation energy. Studies have implicated that antibacterial flavonoids interfere with bacterial quorum sensing through multiple ways including selective targeting of bacterial cells and the inhibition of virulence factors preventing the formation of biofilms [117]. Research into cotton wilt resistance highlighted the role of flavonoids, where metabolomics and transcriptomic analyses revealed enriched flavonoid levels in leaves due to upregulated biosynthesis genes. In red cotton cultivars, heightened flavonoid levels suppressed fungal pathogen invasion by *Verticillium dahliae* [118]. Metabolomics also indicated that plants combatting *Colletotrichum gloeosporioides* infection accumulate flavonoids, suggesting these compounds as potential targets for enhancing resistance through breeding [119]. Flavonoids are classified into several classes such as anthocyanins, flavonols, flavones, flavanones, dihydroflavonols, aurones, chalcones, flavans, and proanthocyanidins [120]. With over 5,000 types identified in plants, specific flavonoids like flavones, flavonols, flavans, flavan 3-ols, proanthocyanidins, flavonones, and isoflavonoids have been investigated for their use in pest management as feeding deterrents [121]. For instance, compounds such as flavones including 5-hydroxyisoderricin, 5-methoxyisoronchocarpin and 7-methoxy-8-(3-methylbutadienyl)-flavanone from *Tephrosia* species have shown deterrent effects against *Spodoptera exempta* and *S. littoralis* [121]. Enhancing flavonoid production by overexpressing specific transcription factors in *Arabidopsis* has been shown to confer resistance against *S. frugiperda* [122]. In addition, a wide range of isoflavonoids like angustone A, licoisoflavone B, angustone B, and angustone C, as well as licoisoflavone A, luteone, licoisoflavone B, and wighteone, have been shown to exhibit antifungal activity against *Colletotrichum gloeosporioides* and *Cladosporium cladosporioides* in addition to deterrence against feeding by insects [123].

Role of Lignin

Lignin is a phenolic heteropolymer found to be essential in plant's defence against insects and pathogens. It restricts pathogen access by physically obstructing or increasing leaf hardness, which diminishes herbivore eating [101]. Due to herbivory or pathogenic attack lignin synthesis is induced followed by its fast deposition which helps to reduce pathogen or herbivore fecundity [124]. It has been reported that plants afflicted with pests and diseases show higher expression levels of lignin-associated genes such as CAD/CAD-like genes [125]. Quinones produced by the oxidation of phenols show covalent interactions with leaf proteins and block them. In addition, quinones show direct toxicity towards insects [126]. Phenols activate defense enzymes *via* the reduction of reactive oxygen species (ROS to superoxide anion and hydroxide radicals, H_2O_2, and singlet oxygen), which trigger a cascade of processes that culminate in enzyme activation [127]. Enhanced deposition of lignin as a response to infection has been reported in wheat infected with spot blotch pathogen [128]. Similar findings have been revealed in Olive plantations tolerant to "Olive Quick Decline Syndrome" as a mitigation measure against X*ylella fastidiosa* [129].

Phytoalexins and Phytoanticipins-mediated Elicitation

Plants employ active defense mechanisms that involve small antimicrobial compounds called phytoalexins, which are synthesized in response to biotic stress or infection [130]. Phytoalexins bear specificity to various plant groups, for instance, isoflavonoids occur in most legumes, terpenoids are abundant in the Solanaceae family [131], crucifers possess glucosinolates, beans, and peas have Phaseollin and pisatin, respectively while as glyceollin is found in alfalfa, soybeans and clover. Examples of well-studied phytoalexins include rishitin in potatoes, capsidiol in peppers, and gossypol in cotton [132]. Phytoanticipins, another class of plant defenses, are antimicrobial compounds present in plant tissues before infection occurs [133]. Saponins are a type of phytoanticipin found in many plants, with notable examples like avenacine in oats and α-tomatine in tomatoes, which have been extensively researched [134].

HERBIVORY

Plants allocate energy and nutrients to grow vital tissues like roots, stems, leaves, and reproductive organs. Herbivore consumption of these tissues can hinder the plant's ability to channel resources into reproduction. Natural selection therefore favors plant traits that mitigate herbivory's negative impact. Some plants tolerate herbivory by rapidly regenerating tissues, occasionally even enhancing reproduction with mild damage [135]. Conversely, other plants evolve traits that deter herbivore consumption, known as resistance. Herbivores, dependent on plants for sustenance, drive natural selection towards those that can overcome plant defenses, spurring plants to develop new forms of resistance. This ongoing coevolution has led to the development of an array of resistance traits in terrestrial plants [136], encompassing reduced visibility to herbivores, as well as the advancement of structural, chemical, and other indirect defense mechanisms. The chemical defense measures adopted by plants to evade herbivory have been discussed below in detail.

Mechanisms Involving Herbivore-induced Phytochemical Defenses

Once stress signals are detected in plant cells, extensive genomic modifications are initiated. Specific promoters activate various signaling pathways to counteract the induced stress. Biotic stress, triggered by molecules like microbe-associated molecular patterns (MAMPS) or pathogen-associated molecular patterns (PAMPS) bind to the pattern recognition receptors (PRRs), initiating PAMPs-triggered immunity or MAMPs-triggered immunity. This cascade involves calcium influx and generation of reactive oxygen species [137]. Salicylic acid also ROS levels to mitigate the cellular oxidative stress [138]. Its synergistic action with other plant hormones contributes significantly to stress tolerance [139]. Moreover, salicylic acid enables plants to resist against the disease-causing microbes and other pathogenic elements [140]. Furthermore, its involvement in the synthesis and accumulation of plant secondary metabolites has also been implicated [141]. Besides, other signalling cascades, the MAPK (mitogen-activated protein kinase) pathways are also integral in responding to biotic stress [142]. In addition, the jasmonic acid signaling pathway also aids plants in coping with stressful conditions, facilitating the production of secondary metabolites such as terpenoids, phytoalexins and alkaloids known for their antimicrobial properties [143].

ROLE OF SECONDARY METABOLITES IN DETERRING HERBIVORES

A variety of chemical substances that enable plants to evade herbivory and provide resistance against different kinds of pathogenic elements can be classified as follows:

Feeding and Oviposition Deterrents

Antifeedants are substances that deter insects from feeding without causing their immediate death; instead, insects cease feeding and eventually starve. These compounds are considered environmentally friendly alternatives for pest control in both agricultural and storage contexts. The exact mechanisms by which antifeedants work are not fully understood, but they are thought to disrupt insect feeding behavior through chemosensory-based aversion to food sources [144] For instance, research by Lin *et al.*, has identified several compounds within the Meliaceae family, including azadirone, 7-deacetyl gedunin gedunin, salannol, methyl angolensate, salannin, azadirachtin and azadiradione as effective deterrents against various insects, particularly lepidopterans [145, 146]. Similarly, several kinds of limonoids derived from *Croton jatrophoides* have demonstrated antifeeding properties against a wide range of feeding insects [147]. Pyrethrins also exhibit similar deterrent effects against various sucking insects [148]. In addition to feeding deterrents, plants release metabolites referred to as oviposition deterrents to prevent the laying of insect eggs, which upon hatching into larvae pose a serious threat to plants [149]. Compounds like coumarin and rutin found in cabbage have been identified as effective oviposition deterrents. The ginsenosides found in *Panax ginseng* plants have been found to affect the feeding behavior and oviposition activities of *Pieris rapae* butterflies [150]. Acyl sugars have also been highlighted by Lin *et al.* [145] as effective oviposition deterrents against various thrips species, including *Frankliniella fusca* and *Frankliniella occidentalis*, particularly in Solanaceae plants [151].

Competition

Production of Phytochemicals: Allelopathy and Competitive Interactions

Plants have the ability to adjust their physical and biochemical characteristics based on environmental conditions, especially to produce secondary metabolites [152]. Allelopathic effects are influenced by the synthesis and release of allelochemicals, making allelopathy inherently responsive to environmental factors. These interactions primarily occur between plants of the same species and different species when they grow together, often in competitive situations. Besides interplant competition, plants also release allelochemicals as a defense mechanism against herbivores, pathogens, insects, and harmful microorganisms. Allelochemicals play diverse roles in plants' defenses against herbivores, insects, and pathogens [153]. Allelopathy in plants can be triggered by internal signals or external influences. These factors amplify the impact of allelopathy, encompassing both biotic and abiotic elements. Biotic factors involve plant competition [154], feeding by animals or insects [155] and microbial interactions [156]. Allelopathy plays a crucial role in plant defense mechanisms, involving the release of various low-molecular-weight compounds that affect their environment by inhibiting the growth or development of nearby plants [157]. Studies have shown that allelochemicals like flavonoids, phenoxazinones, phenolic acids, and benzoxazinones from different wheat genotypes can inhibit weed species such as *Avena fatua, Chenopodium album, Portulaca oleracea, Bromus japonicus,* and *Lolium rigidum* [158]. Similarly, cumin aldehyde derived from *Cuminum cyminum* L. has been found to suppress the growth of both monocot and dicot plants [159]. Avenicin, produced by oats, is another important allelochemical known for its potent activity against plant pathogens [157]. Additionally, secondary metabolites like procyanidins released by *Fallopia* spp. have been shown to reduce the availability of nitrogen in the rhizosphere of invading plants, thereby impacting the growth and nitrogen availability for the invasive community [157, 160]. Plants utilize volatile organic compounds as a form of induced indirect defense to attract natural predators of herbivores [161]. These herbivore-induced plant volatiles (HIPVs) primarily include terpenoids, phenylpropanoids/benzenoids, fatty acids derivates, and several amino acids [162]. For instance, Microplitis mediator, an insect predator, is attracted to HIPVs emitted by cotton plants infected with *Agrotis segetum*, demonstrating the effectiveness of this indirect defense mechanism [154]. Similarly, cucumber and potato plants infested with aphids or caterpillars release HIPVs that serve as cues for ants [163]. *Arabidopsis* plants infested with aphids and caterpillars also emit volatiles that attract parasitoids like *Diadegma semiclausum*, which are significant natural enemies of herbivorous insects [164]. Additionally, plants possess extra floral nectaries containing secondary metabolites that attract ants, which in turn contribute to reducing herbivore attacks [165].

FUTURE PERSPECTIVES

As we delve into the intricate relationship between environmental factors and phytochemical production, the field of phytochemistry is undergoing rapid evolution. Emerging trends and technologies, combined with a deeper understanding of ecological dynamics, are fostering innovative strategies to bolster plant defense mechanisms through phytochemical enhancement. Biotechnological advancements such as synthetic biology and CRISPR-Cas9 enable precise manipulation of metabolic pathways to elevate phytochemical synthesis. Omics technologies—including genomics, proteomics, metabolomics, and transcriptomics—offer crucial insights into regulatory networks and the impact of environmental cues on phytochemical biosynthesis. Integrating these approaches can pinpoint genetic markers that optimize phytochemical profiles in crops.

Advanced cultivation techniques such as vertical farming and controlled environment agriculture (CEA) optimize phytochemical production by regulating light, temperature, and nutrient availability. Methods like hydroponics and aeroponics enhance nutrient absorption, leading to increased yields. Holistic approaches that merge environmental and genetic strategies are indispensable for sustainable phytochemical fortification. Plant-microbe interactions, including beneficial microbes, bolster plant defenses and augment phytochemical synthesis. Agroecological practices—like crop rotation, intercropping, and cover cropping—foster biodiversity and soil health, supporting robust phytochemical production.

Climate change poses challenges to phytochemical biosynthesis, necessitating research into resilient plant varieties. Collaboration among academia, industry, and policymakers is vital for sharing knowledge and driving innovation. Community engagement promotes sustainable agricultural practices that sustain phytochemical production. Regulatory frameworks must adapt to biotechnological advancements in phytochemical research to ensure safety and environmental protection while encouraging innovation. Policies supporting research and sustainable practices are pivotal for the future of phytochemical production, contributing to healthier plants, ecosystems, and communities.

CONCLUSION

Understanding the intricate relationship between environmental factors and plant life is crucial in environmental phytochemistry. This field investigates how variables such as temperature, sunlight, rainfall, and altitude influence the production of phytochemicals—biologically active compounds essential for pharmaceuticals, nutraceuticals, and herbal remedies. Climate plays a pivotal role, significantly shaping phytochemical profiles. Warmer climates often enhance the production of phenolic compounds, vital for plant defense and medicinal properties, while cooler temperatures alter resource allocation and phytochemical composition. Altitude also influences phytochemical profiles; higher elevations stimulate plants to increase the synthesis of secondary metabolites, boosting their bioactive potency as a defense mechanism. Soil composition and nutrient availability are critical factors; nitrogen-rich soils, for instance, promote higher alkaloid production, offering a spectrum of medicinal benefits. Strategic cultivation practices capitalize on these factors to optimize phytochemical yields. Environmental stressors such as pollution and drought prompt plants to intensify the synthesis of secondary metabolites, enriching their bioactive compounds. This adaptive response not only aids in developing stress-tolerant crops but also identifies valuable medicinal resources. Thus, comprehending how climate, altitude, soil composition, and stressors influence plant biochemistry is essential across scientific disciplines. This knowledge deepens our understanding of plant physiology and provides insights into medicine, agriculture, and ecology. Unraveling these interactions opens avenues for drug discovery and sustainable agricultural practices, advancing scientific progress significantly.

REFERENCES

[1]　Das SK, Patra JK, Thatoi H. Antioxidative response to abiotic and biotic stresses in mangrove plants: A review. Int Rev Hydrobiol 2016; 101(1-2): 3-19.
[http://dx.doi.org/10.1002/iroh.201401744]

[2]　Yousuf P, Razzak S, Parvaiz S, Rather YA, Lone R. Role of plant phenolics in the resistance mechanism of plants against insects. InPlant Phenolics in Biotic Stress Management 2024 Feb 28 (pp. 191-215). Singapore: Springer Nature Singapore.
[http://dx.doi.org/10.1007/978-981-99-3334-1_8]

[3]　Kariñho-Betancourt E. Coevolution: plant-herbivore interactions and secondary metabolites of plants. Co-evolution of secondary metabolites. 2020:47-76.
[http://dx.doi.org/10.1007/978-3-319-96397-6_41]

[4]　Khameneh B, Eskin NAM, Iranshahy M, Fazly Bazzaz BS. Phytochemicals: a promising weapon in the arsenal against antibiotic-resistant bacteria. Antibiotics (Basel) 2021; 10(9): 1044.
[http://dx.doi.org/10.3390/antibiotics10091044] [PMID: 34572626]

[5]　Isah T. Stress and defense responses in plant secondary metabolites production. Biol Res 2019; 52(1): 39.
[http://dx.doi.org/10.1186/s40659-019-0246-3] [PMID: 31358053]

[6]　Link KP, Angell HR, Walker JC. The isolation of protocatechuic acid from pigmented onion scales and its significance in relation to disease resistance in onions. J Biol Chem 1929; 81(2): 369-75.
[http://dx.doi.org/10.1016/S0021-9258(18)83819-4]

[7]　Zaynab M, Fatima M, Abbas S, *et al.* Role of secondary metabolites in plant defense against pathogens. Microb Pathog 2018; 124: 198-202.
[http://dx.doi.org/10.1016/j.micpath.2018.08.034] [PMID: 30145251]

[8]　Humphry M, Bednarek P, Kemmerling B, *et al.* A regulon conserved in monocot and dicot plants defines a functional module in antifungal plant immunity. Proc Natl Acad Sci USA 2010; 107(50): 21896-901.
[http://dx.doi.org/10.1073/pnas.1003619107] [PMID: 21098265]

[9]　Morrissey J, Guerinot ML. Iron uptake and transport in plants: the good, the bad, and the ionome. Chem Rev 2009; 109(10): 4553-67.
[http://dx.doi.org/10.1021/cr900112r] [PMID: 19754138]

[10]　Wink M. Plant secondary metabolism: diversity, function and its evolution. Natural Product Communications. 2008 Aug;3(8):
[http://dx.doi.org/10.1177/1934578X0800300801]

[11]　Hettenhausen C, Schuman MC, Wu J. MAPK signaling: A key element in plant defense response to insects. Insect Sci 2015; 22(2): 157-64.

[http://dx.doi.org/10.1111/1744-7917.12128] [PMID: 24753304]

[12]   Schweiger R, Heise AM, Persicke M, Müller C. Interactions between the jasmonic and salicylic acid pathway modulate the plant metabolome and affect herbivores of different feeding types. Plant Cell Environ 2014; 37(7): 1574-85.
[http://dx.doi.org/10.1111/pce.12257] [PMID: 24372400]

[13]   Mason PA, Singer MS. Defensive mixology: combining acquired chemicals towards defence. Funct Ecol 2015; 29(4): 441-50.
[http://dx.doi.org/10.1111/1365-2435.12380]

[14]   War AR, Buhroo AA, Hussain B, Ahmad T, Nair RM, Sharma HC. Plant defense and insect adaptation with reference to secondary metabolites.I n: J.-M. Mérillon and K. G. Ramawat, Eds., Co-evolution of secondary metabolites. 2020:795-822.
[http://dx.doi.org/10.1007/978-3-319-96397-6_60]

[15]   Ben Mrid R, Benmrid B, Hafsa J, Boukcim H, Sobeh M, Yasri A. Secondary metabolites as biostimulant and bioprotectant agents: A review. Sci Total Environ 2021; 777: 146204.
[http://dx.doi.org/10.1016/j.scitotenv.2021.146204]

[16]   Verma N, Shukla S. Impact of various factors responsible for fluctuation in plant secondary metabolites. J Appl Res Med Aromat Plants 2015; 2(4): 105-13.
[http://dx.doi.org/10.1016/j.jarmap.2015.09.002]

[17]   Barton KE. Early ontogenetic patterns in chemical defense in *Plantago* (Plantaginaceae): genetic variation and trade-offs. Am J Bot 2007; 94(1): 56-66.
[http://dx.doi.org/10.3732/ajb.94.1.56] [PMID: 21642208]

[18]   Broun P, Liu Y, Queen E, Schwarz Y, Abenes ML, Leibman M. Importance of transcription factors in the regulation of plant secondary metabolism and their relevance to the control of terpenoid accumulation. Phytochem Rev 2006; 5(1): 27-38.
[http://dx.doi.org/10.1007/s11101-006-9000-x]

[19]   Gouvea DR, Gobbo-Neto L, Sakamoto HT, *et al.* Seasonal variation of the major secondary metabolites present in the extract of *Eremanthus mattogrossensis* Less (Asteraceae: Vernonieae) leaves. Quim Nova 2012; 35(11): 2139-45.
[http://dx.doi.org/10.1590/S0100-40422012001100007]

[20]   Akula R, Ravishankar GA. Influence of abiotic stress signals on secondary metabolites in plants. Plant Signal Behav 2011; 6(11): 1720-31.
[http://dx.doi.org/10.4161/psb.6.11.17613] [PMID: 22041989]

[21]   Ashraf MA, Iqbal M, Rasheed R, Hussain I, Riaz M, Arif MS. Environmental stress and secondary metabolites in plants: an overview. In: P. Ahmad, M. A. Ahanger, V. P. Singh, D. K. Tripathi, P. Alam, and M. N. Alyemeni, Eds., Plant metabolites and regulation under environmental stress. 2018 Jan 1:153-67.
[http://dx.doi.org/10.1016/B978-0-12-812689-9.00008-X]

[22]   Zhi-lin Y, Chuan-chao D, Lian-qing C. Regulation and accumulation of secondary metabolites in plant-fungus symbiotic system. Afr J Biotechnol 2007; 6(11).

[23]   Radušienė J, Karpavičienė B, Stanius Ž. Effect of external and internal factors on secondary metabolites accumulation in St. John's worth. Bot Lith 2012; 18(2): 101-8.
[http://dx.doi.org/10.2478/v10279-012-0012-8]

[24]   Dixon RA, Paiva NL. Stress-induced phenylpropanoid metabolism. Plant Cell 1995; 7(7): 1085-97.
[http://dx.doi.org/10.2307/3870059] [PMID: 12242399]

[25]   Taiz L, Zeiger E. Secondary metabolites and plant defense. Plant Physiol 2006; 4(315)

[26]   Wojakowska A, Muth D, Narożna D, Mądrzak C, Stobiecki M, Kachlicki P. Changes of phenolic secondary metabolite profiles in the reaction of narrow leaf lupin (*Lupinus angustifolius*) plants to infections with *Colletotrichum lupini* fungus or treatment with its toxin. Metabolomics 2013; 9(3):

575-89.
[http://dx.doi.org/10.1007/s11306-012-0475-8] [PMID: 23678343]

[27]   Masson-Delmotte V, Zhai P, Pirani A, Connors SL, Péan C, Berger S, Caud N, Chen Y, Goldfarb L, Gomis MI, Huang M. Climate change 2021: the physical science basis. Contribution of working group I to the sixth assessment report of the intergovernmental panel on climate change. 2021 Jun; 2(1): 2391.

[28]   Qaderi MM, Kurepin LV, Reid DM. Growth and physiological responses of canola (*Brassica napus*) to three components of global climate change: temperature, carbon dioxide and drought. Physiol Plant 2006; 128(4): 710-21.
[http://dx.doi.org/10.1111/j.1399-3054.2006.00804.x]

[29]   Wahid A. Physiological implications of metabolite biosynthesis for net assimilation and heat-stress tolerance of sugarcane (*Saccharum officinarum*) sprouts. J Plant Res 2007; 120(2): 219-28.
[http://dx.doi.org/10.1007/s10265-006-0040-5] [PMID: 17024517]

[30]   Ohama N, Sato H, Shinozaki K, Yamaguchi-Shinozaki K. Transcriptional regulatory network of plant heat stress response. Trends Plant Sci 2017; 22(1): 53-65.
[http://dx.doi.org/10.1016/j.tplants.2016.08.015] [PMID: 27666516]

[31]   Hancock RD, Morris WL, Ducreux LJM, *et al.* Physiological, biochemical and molecular responses of the potato (*Solanum tuberosum* L.) plant to moderately elevated temperature. Plant Cell Environ 2014; 37(2): 439-50.
[http://dx.doi.org/10.1111/pce.12168] [PMID: 23889235]

[32]   Bahuguna RN, Jagadish KSV. Temperature regulation of plant phenological development. Environ Exp Bot 2015; 111: 83-90.
[http://dx.doi.org/10.1016/j.envexpbot.2014.10.007]

[33]   Kaplan F, Kopka J, Haskell DW, *et al.* Exploring the temperature-stress metabolome of *Arabidopsis.* Plant Physiol 2004; 136(4): 4159-68.
[http://dx.doi.org/10.1104/pp.104.052142] [PMID: 15557093]

[34]   Ribeiro PR, Fernandez LG, de Castro RD, Ligterink W, Hilhorst HWM. Physiological and biochemical responses of *Ricinus communis* seedlings to different temperatures: a metabolomics approach. BMC Plant Biol 2014; 14(1): 223.
[http://dx.doi.org/10.1186/s12870-014-0223-5] [PMID: 25109402]

[35]   Zobayed SMA, Afreen F, Kozai T. Temperature stress can alter the photosynthetic efficiency and secondary metabolite concentrations in St. John's wort. Plant Physiol Biochem 2005; 43(10-11): 977-84.
[http://dx.doi.org/10.1016/j.plaphy.2005.07.013] [PMID: 16310362]

[36]   Schonhof I, Kläring HP, Krumbein A, Claußen W, Schreiner M. Effect of temperature increase under low radiation conditions on phytochemicals and ascorbic acid in greenhouse grown broccoli. Agric Ecosyst Environ 2007; 119(1-2): 103-11.
[http://dx.doi.org/10.1016/j.agee.2006.06.018]

[37]   Shohael AM, Ali MB, Yu KW, Hahn EJ, Islam R, Paek KY. Effect of light on oxidative stress, secondary metabolites and induction of antioxidant enzymes in *Eleutherococcus senticosus* somatic embryos in bioreactor. Process Biochem 2006; 41(5): 1179-85.
[http://dx.doi.org/10.1016/j.procbio.2005.12.015]

[38]   Neugart S, Kläring HP, Zietz M, *et al.* The effect of temperature and radiation on flavonol aglycones and flavonol glycosides of kale (*Brassica oleracea var. sabellica*). Food Chem 2012; 133(4): 1456-65.
[http://dx.doi.org/10.1016/j.foodchem.2012.02.034]

[39]   Cohen SD, Tarara JM, Kennedy JA. Assessing the impact of temperature on grape phenolic metabolism. Anal Chim Acta 2008; 621(1): 57-67.
[http://dx.doi.org/10.1016/j.aca.2007.11.029] [PMID: 18573371]

[40]   Ban Y, Honda C, Hatsuyama Y, Igarashi M, Bessho H, Moriguchi T. Isolation and functional analysis of a MYB transcription factor gene that is a key regulator for the development of red coloration in apple skin. Plant Cell Physiol 2007; 48(7): 958-70.
[http://dx.doi.org/10.1093/pcp/pcm066] [PMID: 17526919]

[41]   Chaves I, Passarinho JAP, Capitão C, Chaves MM, Fevereiro P, Ricardo CPP. Temperature stress effects in *Quercus suber* leaf metabolism. J Plant Physiol 2011; 168(15): 1729-34.
[http://dx.doi.org/10.1016/j.jplph.2011.05.013] [PMID: 21676491]

[42]   Goh HH, Khairudin K, Sukiran NA, Normah MN, Baharum SN. Metabolite profiling reveals temperature effects on the VOCs and flavonoids of different plant populations. Plant Biol 2016; 18(S1) (Suppl. 1): 130-9.
[http://dx.doi.org/10.1111/plb.12403] [PMID: 26417881]

[43]   Rosenfeld HJ, Aaby K, Lea P. Influence of temperature and plant density on sensory quality and volatile terpenoids of carrot (*Daucus carota* L.) root. J Sci Food Agric 2002; 82(12): 1384-90.
[http://dx.doi.org/10.1002/jsfa.1200]

[44]   Brown GD. The biosynthesis of artemisinin (Qinghaosu) and the phytochemistry of *Artemisia annua* L. (Qinghao). Molecules 2010; 15(11): 7603-98.
[http://dx.doi.org/10.3390/molecules15117603] [PMID: 21030913]

[45]   Glaubitz U, Erban A, Kopka J, Hincha DK, Zuther E. High night temperature strongly impacts TCA cycle, amino acid and polyamine biosynthetic pathways in rice in a sensitivity-dependent manner. J Exp Bot 2015; 66(20): 6385-97.
[http://dx.doi.org/10.1093/jxb/erv352] [PMID: 26208642]

[46]   Reardon ME, Qaderi MM. Individual and interactive effects of temperature, carbon dioxide and abscisic acid on mung bean (*Vigna radiata*) plants. J Plant Interact 2017; 12(1): 295-303.
[http://dx.doi.org/10.1080/17429145.2017.1353654]

[47]   Sobuj N, Virjamo V, Zhang Y, Nybakken L, Julkunen-Tiitto R. Impacts of elevated temperature and CO_2 concentration on growth and phenolics in the sexually dimorphic *Populus tremula* (L.). Environ Exp Bot 2018; 146: 34-44.
[http://dx.doi.org/10.1016/j.envexpbot.2017.08.003]

[48]   Kleinwächter M, Selmar D. New insights explain that drought stress enhances the quality of spice and medicinal plants: potential applications. Agron Sustain Dev 2015; 35(1): 121-31.
[http://dx.doi.org/10.1007/s13593-014-0260-3]

[49]   Amiri R, Nikbakht A, Etemadi N. Alleviation of drought stress on rose geranium [*Pelargonium graveolens* (L.) Herit.] in terms of antioxidant activity and secondary metabolites by mycorrhizal inoculation. Sci Hortic (Amsterdam) 2015; 197: 373-80.
[http://dx.doi.org/10.1016/j.scienta.2015.09.062]

[50]   Kleinwächter M, Paulsen J, Bloem E, Schnug E, Selmar D. Moderate drought and signal transducer induced biosynthesis of relevant secondary metabolites in thyme (*Thymus vulgaris*), greater celandine (*Chelidonium majus*) and parsley (*Petroselinum crispum*). Ind Crops Prod 2015; 64: 158-66.
[http://dx.doi.org/10.1016/j.indcrop.2014.10.062]

[51]   Bowne JB, Erwin TA, Juttner J, *et al.* Drought responses of leaf tissues from wheat cultivars of differing drought tolerance at the metabolite level. Mol Plant 2012; 5(2): 418-29.
[http://dx.doi.org/10.1093/mp/ssr114] [PMID: 22207720]

[52]   Chan KX, Wirtz M, Phua SY, Estavillo GM, Pogson BJ. Balancing metabolites in drought: the sulfur assimilation conundrum. Trends Plant Sci 2013; 18(1): 18-29.
[http://dx.doi.org/10.1016/j.tplants.2012.07.005] [PMID: 23040678]

[53]   Hura T, Hura K, Grzesiak S. Contents of total phenolics and ferulic acid, and PAL activity during water potential changes in leaves of maize single-cross hybrids of different drought tolerance. J Agron Crop Sci 2008; 194(2): 104-12.

[http://dx.doi.org/10.1111/j.1439-037X.2008.00297.x]

[54]　Bloem E, Haneklaus S, Kleinwächter M, Paulsen J, Schnug E, Selmar D. Stress-induced changes of bioactive compounds in *Tropaeolum majus* L. Ind Crops Prod 2014; 60: 349-59.
[http://dx.doi.org/10.1016/j.indcrop.2014.06.040]

[55]　Ballhorn DJ, Reisdorff C, Pfanz H. Quantitative effects of enhanced CO_2 on jasmonic acid induced plant volatiles of lima bean (*Phaseolus lunatus* L.). J Appl Bot Food Qual 2011; 84: 65-71.

[56]　Lisar SY, Motafakkerazad R, Hossain MM, Rahman IM. Water stress in plants: Causes, effects and responses. In: I. M. M. Rahman and H. Hasegawa Eds., Water stress. 2012 Jan 25; 25(1): 33.

[57]　Azhar NA, Hussain B, Ashraf MY, Abbasi KY. Water stress mediated changes in growth, physiology and secondary metabolites of desi ajwain (*Trachyspermum ammi* L.). Pak J Bot 2011; 43(9): 15-9.

[58]　Zhang L, Wang Q, Guo Q, *et al.* Growth, physiological characteristics and total flavonoid content of *Glechoma longituba* in response to water stress. J Med Plants Res 2012; 6(6): 1015-24.

[59]　Khalid KA. Influence of water stress on growth, essential oil, and chemical composition of herbs. Int Agrophys 2006; 20(4) [*Ocimum* sp.].

[60]　Fazal H, Abbasi BH, Ahmad N, Ali SS, Akbar F, Kanwal F. Correlation of different spectral lights with biomass accumulation and production of antioxidant secondary metabolites in callus cultures of medicinally important *Prunella vulgaris* L. J Photochem Photobiol B 2016; 159: 1-7.
[http://dx.doi.org/10.1016/j.jphotobiol.2016.03.008] [PMID: 26995670]

[61]　Qaderi MM, Martel AB, Strugnell CA. Environmental factors regulate plant secondary metabolites. Plants 2023; 12(3): 447.
[http://dx.doi.org/10.3390/plants12030447] [PMID: 36771531]

[62]　Kopsell DA, Sams CE, Morrow RC. Blue wavelengths from LED lighting increase nutritionally important metabolites in specialty crops. HortScience 2015; 50(9): 1285-8.
[http://dx.doi.org/10.21273/HORTSCI.50.9.1285]

[63]　Darko E, Heydarizadeh P, Schoefs B, Sabzalian MR. Photosynthesis under artificial light: the shift in primary and secondary metabolism. Philos Trans R Soc Lond B Biol Sci 2014; 369(1640): 20130243.
[http://dx.doi.org/10.1098/rstb.2013.0243] [PMID: 24591723]

[64]　Hemm MR, Rider SD, Ogas J, Murry DJ, Chapple C. Light induces phenylpropanoid metabolism in *Arabidopsis* roots. Plant J 2004; 38(5): 765-78.
[http://dx.doi.org/10.1111/j.1365-313X.2004.02089.x] [PMID: 15144378]

[65]　Ahmad N, Rab A, Ahmad N. Light-induced biochemical variations in secondary metabolite production and antioxidant activity in callus cultures of *Stevia rebaudiana* (Bert). J Photochem Photobiol B 2016; 154: 51-6.
[http://dx.doi.org/10.1016/j.jphotobiol.2015.11.015] [PMID: 26688290]

[66]　Ouzounis T, Fretté X, Rosenqvist E, Ottosen CO. Spectral effects of supplementary lighting on the secondary metabolites in roses, chrysanthemums, and campanulas. J Plant Physiol 2014; 171(16): 1491-9.
[http://dx.doi.org/10.1016/j.jplph.2014.06.012] [PMID: 25105234]

[67]　Tattini M, Di Ferdinando M, Brunetti C, *et al.* Esculetin and esculin (esculetin 6-O-glucoside) occur as inclusions and are differentially distributed in the vacuole of palisade cells in *Fraxinus ornus* leaves: A fluorescence microscopy analysis. J Photochem Photobiol B 2014; 140: 28-35.
[http://dx.doi.org/10.1016/j.jphotobiol.2014.06.012] [PMID: 25063983]

[68]　Dao L, Friedman M. Chlorophyll, chlorogenic acid, glycoalkaloid, and protease inhibitor content of fresh and green potatoes. J Agric Food Chem 1994; 42(3): 633-9.
[http://dx.doi.org/10.1021/jf00039a006]

[69]　Ibrahim MH, Jaafar HZE. Primary, secondary metabolites, H_2O_2, malondialdehyde and photosynthetic responses of *Orthosiphon stimaneus* Benth. to different irradiance levels. Molecules 2012; 17(2):

1159-76.
[http://dx.doi.org/10.3390/molecules17021159] [PMID: 22286668]

[70]   Estell RE, Fredrickson EL, James DK. Effect of light intensity and wavelength on concentration of plant secondary metabolites in the leaves of *Flourensia cernua*. Biochem Syst Ecol 2016; 65: 108-14.
[http://dx.doi.org/10.1016/j.bse.2016.02.019]

[71]   Shiga T, Shoji K, Shimada H, Hashida S, Goto F, Yoshihara T. Effect of light quality on rosmarinic acid content and antioxidant activity of sweet basil, *Ocimum basilicum* L. Plant Biotechnol (Tsukuba) 2009; 26(2): 255-9.
[http://dx.doi.org/10.5511/plantbiotechnology.26.255]

[72]   Brazaitytė A, Sakalauskienė S, Samuolienė G, *et al.* The effects of LED illumination spectra and intensity on carotenoid content in *Brassicaceae microgreens*. Food Chem 2015; 173: 600-6.
[http://dx.doi.org/10.1016/j.foodchem.2014.10.077] [PMID: 25466065]

[73]   Charron CS, Sams CE. Glucosinolate content and myrosinase activity in rapid-cycling *Brassica oleracea* grown in a controlled environment. J Am Soc Hortic Sci 2004; 129(3): 321-30.
[http://dx.doi.org/10.21273/JASHS.129.3.0321]

[74]   Keskitalo M. Effect of abiotic growth factors on the concentration of health-promoting secondary metabolites in crops grown in northern latitudes. In: Pfannhauser W, Fenwick GR, Khokar S, editors. Biologically-active phytochemicals in food: analysis, metabolism, bioavailability and function. Cambridge (UK): Royal Society of Chemistry; 2001. p. 34–5.

[75]   Fournier AR, Proctor JTA, Gauthier L, *et al.* Understory light and root ginsenosides in forest-grown *Panax quinquefolius*. Phytochemistry 2003; 63(7): 777-82.
[http://dx.doi.org/10.1016/S0031-9422(03)00346-7] [PMID: 12877918]

[76]   Grunenfelder L, Hiller LK, Knowles NR. Color indices for the assessment of chlorophyll development and greening of fresh market potatoes. Postharvest Biol Technol 2006; 40(1): 73-81.
[http://dx.doi.org/10.1016/j.postharvbio.2005.12.018]

[77]   Engelen-Eigles G, Holden G, Cohen JD, Gardner G. The effect of temperature, photoperiod, and light quality on gluconasturtiin concentration in watercress (*Nasturtium officinale* R. Br.). J Agric Food Chem 2006; 54(2): 328-34.
[http://dx.doi.org/10.1021/jf051857o] [PMID: 16417287]

[78]   Pinto JE, Bertolucci SK, Malta MR, Cardoso MD, de MSilva FA. Coumarin contents in young *Mikania glomerata* plants (Guaco) under different radiation levels and photoperiod. Lat Am J Pharm 2007; 25(3): 387.

[79]   Ashraf M, Iqbal M, Hussain I, Rasheed R. Physiological and biochemical approaches for salinity tolerance. Managing salt tolerance in plants: molecular and genomic perspectives. 2015 Oct 5;79.

[80]   Said-Al Ahl HA, Omer EA. Medicinal and aromatic plants production under salt stress. A review. Herba Pol 2011; 57(2).

[81]   Kováčik J, Klejdus B, Hedbavny J, Bačkor M. Salicylic acid alleviates NaCl-induced changes in the metabolism of *Matricaria chamomilla* plants. Ecotoxicology 2009; 18(5): 544-54.
[http://dx.doi.org/10.1007/s10646-009-0312-7] [PMID: 19381803]

[82]   Jaleel CA. Changes in non enzymatic antioxidants and ajmalicine production in *Catharanthus roseus* with different soil salinity regimes. Molecules 2009; 2: 2.

[83]   Haghighi Z, Karimi N, Modarresi M, Mollayi S. Enhancement of compatible solute and secondary metabolites production in *Plantago ovata* Forsk. by salinity stress. J Med Plants Res. 2012, 16; 6(18): 3495–500.

[84]   Yang L, Wen KS, Ruan X, Zhao YX, Wei F, Wang Q. Response of plant secondary metabolites to environmental factors. Molecules 2018; 23(4): 762.
[http://dx.doi.org/10.3390/molecules23040762] [PMID: 29584636]

[85]   Stewart AJ, Chapman W, Jenkins GI, Graham I, Martin T, Crozier A. The effect of nitrogen and phosphorus deficiency on flavonol accumulation in plant tissues. Plant Cell Environ 2001; 24(11): 1189-97.
[http://dx.doi.org/10.1046/j.1365-3040.2001.00768.x]

[86]   Ibrahim MH, Jaafar HZE, Rahmat A, Rahman ZA. The relationship between phenolics and flavonoids production with total non structural carbohydrate and photosynthetic rate in *Labisia pumila* Benth. under high CO_2 and nitrogen fertilization. Molecules 2010; 16(1): 162-74.
[http://dx.doi.org/10.3390/molecules16010162] [PMID: 21191319]

[87]   Gremigni P, Wong MTF, Edwards NK, Harris D, Hamblin J. Potassium nutrition effects on seed alkaloid concentrations, yield and mineral content of lupins (*Lupinus angustifolius*). Plant Soil 2001; 234(1): 131-42.
[http://dx.doi.org/10.1023/A:1010576702139]

[88]   Lerdau M, Matson P, Fall R, Monson R. Ecological controls over monoterpene emissions from Douglas-fir (*Pseudotsuga menziesii*). Ecology 1995; 76(8): 2640-7.
[http://dx.doi.org/10.2307/2265834]

[89]   Burney OT, Davis AS, Jacobs DF. Phenology of foliar and volatile terpenoid production for *Thuja plicata* families under differential nutrient availability. Environ Exp Bot 2012; 77: 44-52.
[http://dx.doi.org/10.1016/j.envexpbot.2011.11.002]

[90]   Fich EA, Segerson NA, Rose JKC. The plant polyester cutin: biosynthesis, structure, and biological roles. Annu Rev Plant Biol 2016; 67(1): 207-33.
[http://dx.doi.org/10.1146/annurev-arplant-043015-111929] [PMID: 26865339]

[91]   Al-Khayri JM, Rashmi R, Toppo V, *et al.* Plant secondary metabolites: The weapons for biotic stress management. Metabolites 2023; 13(6): 716.
[http://dx.doi.org/10.3390/metabo13060716] [PMID: 37367873]

[92]   Iqbal Z, Iqbal MS, Hashem A, Abd Allah EF, Ansari MI. Plant defense responses to biotic stress and its interplay with fluctuating dark/light conditions. Front Plant Sci 2021; 12: 631810.
[http://dx.doi.org/10.3389/fpls.2021.631810] [PMID: 33763093]

[93]   Sobiczewski P, Iakimova ET, Mikiciński A, Węgrzynowicz-Lesiak E, Dyki B. Necrotrophic behaviour of *Erwinia amylovora* in apple and tobacco leaf tissue. Plant Pathol 2017; 66(5): 842-55.
[http://dx.doi.org/10.1111/ppa.12631]

[94]   Osman HA, Ameen HH, Mohamed M, Elkelany US. Efficacy of integrated microorganisms in controlling root-knot nematode *Meloidogyne javanica* infecting peanut plants under field conditions. Bull Natl Res Cent 2020; 44(1): 134.
[http://dx.doi.org/10.1186/s42269-020-00366-0]

[95]   Pallas V, García JA. How do plant viruses induce disease? Interactions and interference with host components. J Gen Virol 2011; 92(12): 2691-705.
[http://dx.doi.org/10.1099/vir.0.034603-0] [PMID: 21900418]

[96]   Saijo Y, Loo EP. Plant immunity in signal integration between biotic and abiotic stress responses. New Phytol 2020; 225(1): 87-104.
[http://dx.doi.org/10.1111/nph.15989] [PMID: 31209880]

[97]   Howe GA, Jander G. Plant immunity to insect herbivores. Annu Rev Plant Biol 2008; 59(1): 41-66.
[http://dx.doi.org/10.1146/annurev.arplant.59.032607.092825] [PMID: 18031220]

[98]   Morant AV, Jørgensen K, Jørgensen C, *et al.* β-Glucosidases as detonators of plant chemical defense. Phytochemistry 2008; 69(9): 1795-813.
[http://dx.doi.org/10.1016/j.phytochem.2008.03.006] [PMID: 18472115]

[99]   Walling LL. The myriad plant responses to herbivores. J Plant Growth Regul 2000; 19(2): 195-216.
[http://dx.doi.org/10.1007/s003440000026] [PMID: 11038228]

[100] Lamers J, van der Meer T, Testerink C. How plants sense and respond to stressful environments. Plant Physiol 2020; 182(4): 1624-35.
[http://dx.doi.org/10.1104/pp.19.01464] [PMID: 32132112]

[101] War AR, Paulraj MG, Ahmad T, *et al.* Mechanisms of plant defense against insect herbivores. Plant Signal Behav 2012; 7(10): 1306-20.
[http://dx.doi.org/10.4161/psb.21663] [PMID: 22895106]

[102] Gish M, Mescher MC, De Moraes CM. Mechanical defenses of plant extrafloral nectaries against herbivory. Commun Integr Biol 2016; 9(3): e1178431.
[http://dx.doi.org/10.1080/19420889.2016.1178431] [PMID: 27489584]

[103] Moustafa-Farag M, Almoneafy A, Mahmoud A, *et al.* Melatonin and its protective role against biotic stress impacts on plants. Biomolecules 2019; 10(1): 54.
[http://dx.doi.org/10.3390/biom10010054] [PMID: 31905696]

[104] Murch SJ, Erland LAE. A systematic review of melatonin in plants: an example of evolution of literature. Front Plant Sci 2021; 12: 683047.
[http://dx.doi.org/10.3389/fpls.2021.683047] [PMID: 34249052]

[105] Mandal MK, Suren H, Ward B, Boroujerdi A, Kousik C. Differential roles of melatonin in plant-host resistance and pathogen suppression in cucurbits. J Pineal Res 2018; 65(3): e12505.
[http://dx.doi.org/10.1111/jpi.12505] [PMID: 29766569]

[106] Qian Y, Tan DX, Reiter RJ, Shi H. Comparative metabolomic analysis highlights the involvement of sugars and glycerol in melatonin-mediated innate immunity against bacterial pathogen in *Arabidopsis*. Sci Rep 2015; 5(1): 15815.
[http://dx.doi.org/10.1038/srep15815] [PMID: 26508076]

[107] Nehela Y, Killiny N. Infection with phytopathogenic bacterium inhibits melatonin biosynthesis, decreases longevity of its vector, and suppresses the free radical-defense. J Pineal Res 2018; 65(3): e12511.
[http://dx.doi.org/10.1111/jpi.12511] [PMID: 29786865]

[108] Jiao J, Ma Y, Chen S, *et al.* Melatonin-producing endophytic bacteria from grapevine roots promote the abiotic stress-induced production of endogenous melatonin in their hosts. Front Plant Sci 2016; 7: 1387.
[http://dx.doi.org/10.3389/fpls.2016.01387] [PMID: 27708652]

[109] Zhang X, Zhang H, Zhang H, Tang M. Exogenous melatonin application enhances *Rhizophagus irregularis* symbiosis and induces the antioxidant response of *Medicago truncatula* under lead stress. Front Microbiol 2020; 11: 516.
[http://dx.doi.org/10.3389/fmicb.2020.00516] [PMID: 32351459]

[110] Sherif SM, Erland LA, Shukla MR, Saxena PK. Bark and wood tissues of American elm exhibit distinct responses to Dutch elm disease. Sci Rep 2017; 7(1): 7114.
[http://dx.doi.org/10.1038/s41598-017-07779-4] [PMID: 28769110]

[111] Saremba BM, Tymm FJM, Baethke K, *et al.* Plant signals during beetle (*Scolytus multistriatus*) feeding in American elm (*Ulmus americana* Planch). Plant Signal Behav 2017; 12(5): e1296997.
[http://dx.doi.org/10.1080/15592324.2017.1296997] [PMID: 28448744]

[112] Fujiwara T, Maisonneuve S, Isshiki M, *et al.* Sekiguchi lesion gene encodes a cytochrome P450 monooxygenase that catalyzes conversion of tryptamine to serotonin in rice. J Biol Chem 2010; 285(15): 11308-13.
[http://dx.doi.org/10.1074/jbc.M109.091371] [PMID: 20150424]

[113] Lee HY, Byeon Y, Back K. Melatonin as a signal molecule triggering defense responses against pathogen attack in *Arabidopsis* and tobacco. J Pineal Res 2014; 57(3): 262-8.
[http://dx.doi.org/10.1111/jpi.12165] [PMID: 25099383]

[114] Lee HY, Byeon Y, Tan DX, Reiter RJ, Back K. *Arabidopsis* serotonin *N* -acetyltransferase knockout

mutant plants exhibit decreased melatonin and salicylic acid levels resulting in susceptibility to an avirulent pathogen. J Pineal Res 2015; 58(3): 291-9.
[http://dx.doi.org/10.1111/jpi.12214] [PMID: 25652756]

[115] Zhu Z, Lee B. Friends or foes: new insights in jasmonate and ethylene co-actions. Plant Cell Physiol 2015; 56(3): 414-20.
[http://dx.doi.org/10.1093/pcp/pcu171] [PMID: 25435545]

[116] Kaur S, Samota MK, Choudhary M, *et al.* How do plants defend themselves against pathogens-Biochemical mechanisms and genetic interventions. Physiol Mol Biol Plants 2022; 28(2): 485-504.
[http://dx.doi.org/10.1007/s12298-022-01146-y] [PMID: 35400890]

[117] Górniak I, Bartoszewski R, Króliczewski J. Comprehensive review of antimicrobial activities of plant flavonoids. Phytochem Rev 2019; 18(1): 241-72.
[http://dx.doi.org/10.1007/s11101-018-9591-z]

[118] Long L, Liu J, Gao Y, *et al.* Flavonoid accumulation in spontaneous cotton mutant results in red coloration and enhanced disease resistance. Plant Physiol Biochem 2019; 143: 40-9.
[http://dx.doi.org/10.1016/j.plaphy.2019.08.021] [PMID: 31479881]

[119] Jiang L, Wu P, Yang L, *et al.* Transcriptomics and metabolomics reveal the induction of flavonoid biosynthesis pathway in the interaction of Stylosanthes-Colletotrichum gloeosporioides. Genomics 2021; 113(4): 2702-16.
[http://dx.doi.org/10.1016/j.ygeno.2021.06.004] [PMID: 34111523]

[120] Treutter D. Significance of flavonoids in plant resistance: a review. Environ Chem Lett 2006; 4(3): 147-57.
[http://dx.doi.org/10.1007/s10311-006-0068-8]

[121] Simmonds MSJ, Blaney WM, Fellows LE. Behavioral and electrophysiological study of antifeedant mechanisms associated with polyhydroxy alkaloids. J Chem Ecol 1990; 16(11): 3167-96.
[http://dx.doi.org/10.1007/BF00979618] [PMID: 24263302]

[122] Johnson ET, Dowd PF. Differentially enhanced insect resistance, at a cost, in *Arabidopsis thaliana* constitutively expressing a transcription factor of defensive metabolites. J Agric Food Chem 2004; 52(16): 5135-8.
[http://dx.doi.org/10.1021/jf0308049] [PMID: 15291486]

[123] Lane GA, Sutherland ORW, Skipp RA. Isoflavonoids as insect feeding deterrents and antifungal components from root of *Lupinus angustifolius*. J Chem Ecol 1987; 13(4): 771-83.
[http://dx.doi.org/10.1007/BF01020159] [PMID: 24302045]

[124] Johnson MTJ, Smith SD, Rausher MD. Plant sex and the evolution of plant defenses against herbivores. Proc Natl Acad Sci USA 2009; 106(43): 18079-84.
[http://dx.doi.org/10.1073/pnas.0904695106] [PMID: 19617572]

[125] Barakat A, Bagniewska-Zadworna A, Frost CJ, Carlson JE. Phylogeny and expression profiling of CAD and CAD-like genes in hybrid *Populus* (*P. deltoides* × *P. nigra*): evidence from herbivore damage for subfunctionalization and functional divergence. BMC Plant Biol 2010; 10(1): 100.
[http://dx.doi.org/10.1186/1471-2229-10-100] [PMID: 20509918]

[126] Bhonwong A, Stout MJ, Attajarusit J, Tantasawat P. Defensive role of tomato polyphenol oxidases against cotton bollworm (*Helicoverpa armigera*) and beet armyworm (*Spodoptera exigua*). J Chem Ecol 2009; 35(1): 28-38.
[http://dx.doi.org/10.1007/s10886-008-9571-7] [PMID: 19050959]

[127] Maffei ME, Mithöfer A, Boland W. Insects feeding on plants: Rapid signals and responses preceding the induction of phytochemical release. Phytochemistry 2007; 68(22-24): 2946-59.
[http://dx.doi.org/10.1016/j.phytochem.2007.07.016] [PMID: 17825328]

[128] Eisa M, Chand R, Joshi AK. Biochemical and histochemical traits: a promising way to screen resistance against spot blotch (*Bipolaris sorokiniana*) of wheat. Eur J Plant Pathol 2013; 137(4): 805-

20.
[http://dx.doi.org/10.1007/s10658-013-0290-8]

[129]   Sabella E, Luvisi A, Aprile A, *et al. Xylella fastidiosa* induces differential expression of lignification related-genes and lignin accumulation in tolerant olive trees cv. Leccino. J Plant Physiol 2018; 220: 60-8.
[http://dx.doi.org/10.1016/j.jplph.2017.10.007] [PMID: 29149645]

[130]   Wu J. [The "chemical defense" of plants against pathogenic microbes: Phytoalexins biosynthesis and molecular regulations]. Ying Yong Sheng Tai Xue Bao 2020; 31(7): 2161-7.
[PMID: 32715677]

[131]   Agrios GN. How plants defend themselves against pathogens. Plant Pathology 2005; 207-48.
[http://dx.doi.org/10.1016/B978-0-08-047378-9.50012-9]

[132]   Shin M, Umezawa C, Shin T. Natural anti-microbial systems: antimicrobial compounds in plants. In: Batt CA, Tortorello ML, Eds. Encyclopedia of Food Microbiology (Second Edition) 2014,920-929.
[http://dx.doi.org/10.1016/B978-0-12-384730-0.00239-1]

[133]   Oros G, Kállai Z. Phytoanticipins: The constitutive defense compounds as potential botanical fungicides. In: Abdelrahman M, Jogaiah S Eds., Bioactive molecules in plant defense: Signaling in growth and stress. 2019: 179-229.

[134]   González-Lamothe R, Mitchell G, Gattuso M, Diarra MS, Malouin F, Bouarab K. Plant antimicrobial agents and their effects on plant and human pathogens. Int J Mol Sci 2009; 10(8): 3400-19.
[http://dx.doi.org/10.3390/ijms10083400] [PMID: 20111686]

[135]   Stowe KA, Marquis RJ, Hochwender CG, Simms EL. The evolutionary ecology of tolerance to consumer damage. Annu Rev Ecol Syst 2000; 31(1): 565-95.
[http://dx.doi.org/10.1146/annurev.ecolsys.31.1.565]

[136]   Agrawal AA, Fishbein M, Halitschke R, Hastings AP, Rabosky DL, Rasmann S. Evidence for adaptive radiation from a phylogenetic study of plant defenses. Proc Natl Acad Sci USA 2009; 106(43): 18067-72.
[http://dx.doi.org/10.1073/pnas.0904862106] [PMID: 19805160]

[137]   Bhar A, Chakraborty A, Roy A. Plant responses to biotic stress: Old memories matter. Plants 2021; 11(1): 84.
[http://dx.doi.org/10.3390/plants11010084] [PMID: 35009087]

[138]   Lukan T, Coll A. Intertwined roles of reactive oxygen species and salicylic acid signaling are crucial for the plant response to biotic stress. Int J Mol Sci 2022; 23(10): 5568.
[http://dx.doi.org/10.3390/ijms23105568] [PMID: 35628379]

[139]   Koo YM, Heo AY, Choi HW. Salicylic acid as a safe plant protector and growth regulator. Plant Pathol J 2020; 36(1): 1-10.
[http://dx.doi.org/10.5423/PPJ.RW.12.2019.0295] [PMID: 32089657]

[140]   Vlot AC, Dempsey DMA, Klessig DF. Salicylic Acid, a multifaceted hormone to combat disease. Annu Rev Phytopathol 2009; 47(1): 177-206.
[http://dx.doi.org/10.1146/annurev.phyto.050908.135202] [PMID: 19400653]

[141]   Ali B. Salicylic acid: An efficient elicitor of secondary metabolite production in plants. Biocatal Agric Biotechnol 2021; 31: 101884.
[http://dx.doi.org/10.1016/j.bcab.2020.101884]

[142]   Taj G, Agarwal P, Grant M, Kumar A. MAPK machinery in plants. Plant Signal Behav 2010; 5(11): 1370-8.
[http://dx.doi.org/10.4161/psb.5.11.13020] [PMID: 20980831]

[143]   Ghorbel M, Brini F, Sharma A, Landi M. Role of jasmonic acid in plants: the molecular point of view. Plant Cell Rep 2021; 40(8): 1471-94.
[http://dx.doi.org/10.1007/s00299-021-02687-4] [PMID: 33821356]

[144] Szczepanik M, Dams I, Wawrzeńczyk C. Feeding deterrent activity of terpenoid lactones with the p-menthane system against the Colorado potato beetle (Coleoptera: Chrysomelidae). Environ Entomol 2005; 34(6): 1433-40.
[http://dx.doi.org/10.1603/0046-225X-34.6.1433]

[145] Lin M, Yang S, Huang J, Zhou L. Insecticidal triterpenes in Meliaceae: Plant species, molecules and activities: Part I (*Aphanamixis-Chukrasia*). Int J Mol Sci 2021; 22(24): 13262.
[http://dx.doi.org/10.3390/ijms222413262] [PMID: 34948062]

[146] Senthil-Nathan S. Physiological and biochemical effect of neem and other Meliaceae plants secondary metabolites against Lepidopteran insects. Front Physiol 2013; 4: 359.
[http://dx.doi.org/10.3389/fphys.2013.00359] [PMID: 24391591]

[147] Nihei K, Asaka Y, Mine Y, *et al.* Musidunin and musiduol, insect antifeedants from *Croton jatrophoides*. J Nat Prod 2006; 69(6): 975-7.
[http://dx.doi.org/10.1021/np060068d] [PMID: 16792423]

[148] Kojima T, Yamato S, Kawamura S. Natural and synthetic pyrethrins act as feeding deterrents against the black blowfly, *Phormia regina* (Meigen). Insects 2022; 13(8): 678.
[http://dx.doi.org/10.3390/insects13080678] [PMID: 36005302]

[149] Kumari A, Kaushik N. Oviposition deterrents in herbivorous insects and their potential use in integrated pest management. Indian J Exp Biol 2016; 54(3): 163-74.
[PMID: 27145629]

[150] Zhang A, Liu Z, Lei F, *et al.* Antifeedant and oviposition-deterring activity of total ginsenosides against *Pieris rapae*. Saudi J Biol Sci 2017; 24(8): 1751-3.
[http://dx.doi.org/10.1016/j.sjbs.2017.11.005] [PMID: 29551916]

[151] Ben-Mahmoud S, Smeda JR, Chappell TM, *et al.* Acylsugar amount and fatty acid profile differentially suppress oviposition by western flower thrips, *Frankliniella occidentalis*, on tomato and interspecific hybrid flowers. PLoS One 2018; 13(7): e0201583.
[http://dx.doi.org/10.1371/journal.pone.0201583] [PMID: 30063755]

[152] Kong CH, Li Z, Li FL, Xia XX, Wang P. Chemically mediated plant–plant interactions: Allelopathy and allelobiosis. Plants 2024; 13(5): 626.
[http://dx.doi.org/10.3390/plants13050626] [PMID: 38475470]

[153] Shan Z, Zhou S, Shah A, Arafat Y, Arif Hussain Rizvi S, Shao H. Plant allelopathy in response to biotic and abiotic factors. Agronomy (Basel) 2023; 13(9): 2358.
[http://dx.doi.org/10.3390/agronomy13092358]

[154] Li M, Xia S, Zhang T, Williams L III, Xiao H, Lu Y. Volatiles from cotton plants infested by *Agrotis segetum* (Lep.: Noctuidae) attract the larval parasitoid *Microplitis mediator* (Hym.: Braconidae). Plants 2022; 11(7): 863.
[http://dx.doi.org/10.3390/plants11070863] [PMID: 35406842]

[155] Kost C, Heil M. The defensive role of volatile emission and extrafloral nectar secretion for lima bean in nature. J Chem Ecol 2008; 34(1): 2-13.
[http://dx.doi.org/10.1007/s10886-007-9404-0] [PMID: 18071821]

[156] De Lacy Costello BPJ, Evans P, Ewen RJ, *et al.* Gas chromatography–mass spectrometry analyses of volatile organic compounds from potato tubers inoculated with *Phytophthora infestans* or *Fusarium coeruleum*. Plant Pathol 2001; 50(4): 489-96.
[http://dx.doi.org/10.1046/j.1365-3059.2001.00594.x]

[157] Latif S, Chiapusio G, Weston LA. Allelopathy and the role of allelochemicals in plant defence. In Advances in botanical research 2017 Jan 1 (Vol. 82, pp. 19-54). Academic Press.
[http://dx.doi.org/10.1016/bs.abr.2016.12.001]

[158] Hussain MI, Araniti F, Schulz M, *et al.* Benzoxazinoids in wheat allelopathy – From discovery to application for sustainable weed management. Environ Exp Bot 2022; 202: 104997.

[http://dx.doi.org/10.1016/j.envexpbot.2022.104997]

[159]   Sunohara Y, Nakano K, Matsuyama S, Oka T, Matsumoto H. Cuminaldehyde, a cumin seed volatile component, induces growth inhibition, overproduction of reactive oxygen species and cell cycle arrest in onion roots. Sci Hortic (Amsterdam) 2021; 289: 110493.
[http://dx.doi.org/10.1016/j.scienta.2021.110493]

[160]   Bardon C, Piola F, Haichar FZ, *et al.* Identification of B-type procyanidins in *Fallopia* spp. involved in biological denitrification inhibition. Environ Microbiol 2016; 18(2): 644-55.
[http://dx.doi.org/10.1111/1462-2920.13062] [PMID: 26411284]

[161]   Ode PJ. Caterpillars, plant chemistry, and parasitoids in natural vs. agroecosystems. InCaterpillars in the Middle: Tritrophic Interactions in a Changing World 2022 Apr 23 (pp. 395-423). Cham: Springer International Publishing.

[162]   War AR, Sharma HC, Paulraj MG, War MY, Ignacimuthu S. Herbivore induced plant volatiles: Their role in plant defense for pest management. Plant Signal Behav 2011; 6(12): 1973-8.
[http://dx.doi.org/10.4161/psb.6.12.18053] [PMID: 22105032]

[163]   Schettino M, Grasso DA, Weldegergis BT, *et al.* Response of a predatory ant to volatiles emitted by aphid-and caterpillar-infested cucumber and potato plants. J Chem Ecol 2017; 43(10): 1007-22.
[http://dx.doi.org/10.1007/s10886-017-0887-z] [PMID: 28951999]

[164]   Kroes A, Weldegergis BT, Cappai F, Dicke M, van Loon JJA. Terpenoid biosynthesis in *Arabidopsis* attacked by caterpillars and aphids: effects of aphid density on the attraction of a caterpillar parasitoid. Oecologia 2017; 185(4): 699-712.
[http://dx.doi.org/10.1007/s00442-017-3985-2] [PMID: 29052769]

[165]   Cusumano A, Weldegergis BT, Colazza S, Dicke M, Fatouros NE. Attraction of egg-killing parasitoids toward induced plant volatiles in a multi-herbivore context. Oecologia 2015; 179(1): 163-74.
[http://dx.doi.org/10.1007/s00442-015-3325-3] [PMID: 25953114]

CHAPTER 7

Exploiting Phytochemicals for Nematode Management as A Control Strategy

Preety Tomar[1], Gagan Preet Kour Bali[2,*] and Joginder Singh Rilta[3]

[1] *Department of Zoology, Eternal University, Baru Sahib, Sirmaur, Himachal Pradesh, India*

[2] *Department of Biosciences, UIBT, Chandigarh University, Mohali, Punjab, India*

[3] *Department of Biosciences, Himachal Pradesh University, Summer Hill, Shimla, India*

Abstract: Plant-parasitic nematodes, or PPNs, cause significant losses in commercial crops all over the world. Research efforts should be directed toward developing safe and cost-effective control mechanisms due to the health and environmental risks associated with the use of chemical nematicides. An essential component of these initiatives is the wise exploitation of plant-PPN interaction. As research progresses, naturally occurring phytochemicals that are hostile to other nematodes and plant parasites have been discovered. Plants produce a wide range of secondary metabolites that play an excellent role in plant protection. Polythienyls, glucosinolates, isothiocyanates, glycosides, alkaloids, lipids, terpenoids, steroids, triterpenoids, phenolics, and several other classes have been produced by higher plants. This chapter provides insights into the phyto-nematode interactions and production of anti-nematode phytochemicals to protect them from PPNs. Despite being unprofitable in many cases right now, the use of phytochemicals in agriculture has a lot of potential for the future.

Keywords: Alkaloids, Giant cells, Glucoraphanin, Hypersensitivity, Juveniles, Metabolites, Multienzyme, Nematodes, Nematicidal, Plant-parasitic, Resistance, Rhizosphere, Secondary, Stylets, Susceptibility.

INTRODUCTION

The soil region occupying the root area is copious with a broad diversity of microbiomes. The common residents of this area include both plant pathogenic as well as plant beneficial microbiomes [1, 2] Phytoparasitic roundworms causing damage to plants are commonly regarded as plant parasitic nematodes (PPNs) [3]. These PPNs are minute microscopic worms that move around the rhizospheric soil to find their host and finally procure their liquid food material from the plant

---

* **Corresponding author Gagan Preet Kour Bali:** Department of Biosciences, UIBT, Chandigarh University, Mohali, Punjab, India; E-mail: gaganviren@gmail.com

Shivam Jasrotia & Ajay Kumar (Eds.)

roots [4]. They damage the plants, which results in reduced yields or, sometimes, complete plant losses. About 12.6% of crop losses, which account for 216 billion dollars per year, have been predicted by these PPNs throughout the world [5]. Almost all plant species, including all crops, are parasitized by the more than 4100 PPN species that have been identified to date [6].

Different parasitic modes with varied life cycles have been observed in several forms of PPNs [6]. The major disease-causing agents in crops are phytophagous nematodes, which belong to the order Tylenchida [3]. The members of Tylenchida are even pathogenic to several invertebrate species as well as several fungal species [7]. *Pratylenchus, Hoplolaimus, Meloidogyne, Heterodera, Xiphinema,* and *Rotylenchulus* are the most important genera of PPNs responsible for crop damage [8 - 11]. In PPNs, the most flourishing nematodes are sedentary nematodes [12]. The endoparasitic nematode worm, *Pratylenchus penetrans,* is regarded as the causative agent of plant root lesions [13]. They are the active forms and keep on moving through the plant roots in both forms, as adults or juveniles. During the unfavourable conditions inside plant roots, they get away from the roots and roam inside the soil [14]. Other endoparasitic PPNs include the sedentary nematode species *Meloidogyne incognita.* This nematode species is regarded as the root-knot nematode as it forms irregular galls on plant roots. Once the infective stage juvenile (J_2) got into the root, it obtained its diet material near the vascular bundles and underwent development. It then finally loses its moving ability and fixes in one place, forming a gall and completing its life journey inside its roots [11, 15].

The management of these destructive nematodes is quite challenging. Although the use of chemical nematicides has proven effective against these PPNs, they are banned due to their higher cost, low availability, and environmental vulnerability [16 - 18]. In the 20th century, even nematodes have been found to be successful against PPNs, regarded as predatory nematodes. Along with predatory action, they help in nutrient cycling in plants [19]. Recently, more emphasis has been given to the studies of anti-nematode phytochemicals (ANPs) that have been produced by plants to overcome PPN stress [20]. Nematode behaviour, development, reproduction, and survival are influenced by certain plant metabolites secreted from the roots, which minimises plant harm [21, 22]. While there is still little knowledge on how plants alter the nematode community in the rhizosphere, some may even increase nematode-hostile bacteria in the rhizosphere [23, 22]. So, in this chapter, the main emphasis has been placed on the various strategies that plants exhibit (mainly the production of phytochemicals) to defend them from the PPN attack.

PHYTO-NEMATODE INTERACTIONS

A huge variety of crop plants can be completely destroyed by plant-parasitic nematodes, resulting in annual agricultural losses worth billions of dollars [16]. All portions of the plant, including the roots, leaves, stems, flowers, and seeds, are consumed by nematodes. Although nematodes consume plants in a variety of ways, they are all equipped with a unique spear known as a stylet [3]. These phytopathogens drastically alter the appearance and physiology of their hosts [24]. The only food source for all plant parasitic nematodes (PPNs) is the cytoplasm of living plant cells, making them obligatory parasites [25]. Although both cyst and root-knot nematodes engage in intricate interactions with their hosts, their parasitic cycles differ in certain notable ways [26]. After entering roots, cyst nematodes proceed to the vascular cylinder, puncturing cells with their stylets and causing disruptions along the way [27]. They create a feeding site, presumably by injecting stylet secretions. The disintegration of the cell walls separates the original feeding site cell and its surrounding cells, which leads to the growth of a multinucleate syncytium (the establishment of a feeding site) [28]. Before becoming adults, cyst nematodes go through three molts inside the root. They typically procreate sexually, with the female becoming egg-filled after fertilisation [29]. The carcass of the dead female turns into a cyst that shields the eggs.

The juvenile of the root-knot nematode migrates down the plant cortex towards the root tip, moving intracellularly after entering the root [26]. After entering the vascular cylinder's base, the juveniles go up the root [30]. They cause nuclear division in the host cells without cytokinesis, therefore creating a permanent feeding site in the root's differentiation zone [31]. Giant cells, or galls, also known as "root knots," are created when the plant cells surrounding the feeding site proliferate and enlarge [32]. After three moults, the nematodes transform into pear-shaped, egg-laying females by ingesting the cytoplasm of the large cells generated from plants through their stylets [33, 34].

Upon invasion, these nematodes receive their water and nutrition from feeding cells that are induced to redifferentiate by root-knot and cyst nematodes [35]. Plants use a variety of complementary mechanisms to identify PPN invasion, which in turn stimulates immunological responses to PPN infection [36, 37]. Signal transduction, defence gene activation, and other mechanisms comprise the plant's immunological response to parasitic nematodes [38]. Through the use of cell surface-localised pattern recognition receptors (PRRs), plants are able to identify pathogen-associated molecular patterns (PAMPs) formed from PPNs, which results in pattern-triggered immunity (PTI). The recognition of damage-associated molecular patterns (DAMPs) using PRR-based recognition enables plants to identify tissue and cellular damage resulting from PPN invasion or migration. NLR-triggered immunity is a result of resistant plants' additional capacity to identify PPN effectors through intracellular nucleotide-binding domain leucine-rich repeat (NLR)-type immune receptors. Certain PRRs have the ability to identify apoplastic PPN effectors and cause PTI [39, 40].

PRODUCTION OF ANTI-NEMATODE PHYTOCHEMICALS

A diversity of metabolic substances have been produced by plants, and about 200,000 compounds as secondary metabolites have been reported from plant metabolism [20, 41]. Plants produce a wide spectrum of bioactive compounds, such as polythienyls, glucosinolates, isothiocyanates, polyacetylenes, alkaloids, cyanogenic glycosides, lipids, triterpenoids, sesquiterpenoids, terpenoids, steroids, quassinoids, diterpenoids, phenolics, simple and complex [41, 42]. Research have identified several thousands of these chemical compounds, some of which have the ability to inhibit or kill nematodes. Depending on the desired outcomes, plant secondary metabolites can function as nematicidal, nematode repellents, attractants, hatching inhibitors, or boosters [22, 43].

Phenolic Compounds

One of the most common classes of secondary metabolites in the kingdom of plants is phenolics [44]. A benzene ring (C6) and one or more hydroxyl groups characterise this vast and varied class of aromatic chemicals. Phenolic compounds are among the many major classes of allelochemicals known as plant defence compounds (terpenoids, saponins, alkaloids, and glucosinolates) [45]. They have been long linked to plant defence strategies, such as nematode resistance, and their potential to improve both plant and human health has sparked further research interest [46]. Phenolics often consist of simpler monophenol molecules and more complex polyphenol and oligophenol compounds. Plants produce phenolic compounds as secondary metabolites [47]. These phenolic compounds are important for protection against a range of diseases and pests that affect plants [48]. It has been proposed since the early 1960s that phenolic chemicals play a part in nematode resistance. Phenolic compounds are the most potent inhibitors of nematodes and exhibit nematicidal activities (directly or indirectly) by interfering in phyto-nematode interactions [49, 50].

Albuquerque *et al.*, [51] reported that in coffee plants, the cells surrounding the site of feeding of *M. incognita* become dead quickly after the formation of a giant cell due to a hypersensitivity reaction (HR) that resulted in causing resistance in coffee plants against *M. incognita* infection. Benzoic acids, one type of phenolic acid that is exuded into the rhizosphere by soybeans, can function as chemoattractants, facilitating quorum sensing functions among beneficial microbiome [52]. This, in turn, helps protect against infection by nematodes (*M. incognita*) [53]. In a broad range of plant-nematode combinations, it has been observed that increased basal and/or induced quantities of phenolic chemicals are generally correlated with nematode resistance [54, 55]. Yates *et al.*, [56] reported that phenolic acids have strong nematicidal effects *in vitro* and are strongly stimulated during *Heterodera glycines* infection.

Phenolic compounds have been shown to have a direct effect on PPNs [57, 58]. According to Gupta *et al.* [59], 3,4-dimethylphenol exhibited nematicidal action against *M. incognita* under *in vitro* conditions. Similarly, catechol [60], and 2-hydroxinaphthoic acid [61] have all been shown to exhibit nematicidal action. Phenols have also been shown to have a significant role in defense, including resistance against specific species of root-lesion nematodes such as *Pratylenchus coffeae, Pratylenchus penetrans,* and *Pratylenchus zeae* [62 - 64].

Chlorogenic acid, one of the most abundant beneficial polyphenols in plants, is well-known as a nutritional antioxidant included in plant-based diets. Research has shown this molecule to be an excellent defence against a range of insect herbivores [65]. Chemically 5-caffeoylquinic acid is an ester containing both quinic and caffeic acids, referred to as chlorogenic acid. It is plentiful in plants, fruits, and vegetables [66]. Higher concentration of chlorogenic acid has been correlated to nematode resistance in a variety of plant species in areas where PPN infection is present (monocot and dicot plants) [67 - 70]. Additionally, it was recently confirmed that a poplar clone of *Populus tremula* and *Populus alba* strongly inhibits chlorogenic acid accumulation in *M. incognita* galls. This shows that *M. incognita* may cause the inhibition of chlorogenic acid biosynthesis as a pathogenesis strategy [71]. At low doses, *Artemisia annua's* chlorogenic acid and caffeineic acid exhibit strong inhibitory effects against *Globodera rostochiensis* and *Xiphinema index*, while they lack nematicidal activity against *M. incognita* [72]. Mnviajan *et al.* [73], discovered that caffeine acid was extremely harmful to nematode juveniles. Afifah *et al.* [74] reported that in resistant tomato cultivars, caffeic acid accumulates as *M. incognita* infects the tomato.

Glucosinolates

Glucosinolates (GSLs) are hydrophilic secondary metabolites formed from plants, mostly identified in the Brassicales order. They consist of a thioglucose group, a sulfonated oxime group, and a side chain derived from amino acids [75]. Growth-defense balance is one of the physiological processes in which GSLs and the products of their hydrolysis are important [76, 77]. *In vivo*, GLS coexist alongside glycosylated thioglucosidases, or myrosinase(s), which hydrolyze them to produce bioactive isothiocyanates (ITC). According to Avato [78], GLSs and

ITCs serve as defensive bioactive metabolites against herbivores, insects, and plant diseases. There is significant nematicidal activity in GSLs and, in particular, glucosinolate hydrolysis products (GHPs) [79]. These substances have a variety of direct and indirect impacts on nematodes, including an anti-hatching effect, toxicity, and the encouragement of populations of nematophagous bacteria or rival saprophytic worms. GHPs have nematicidal effects on plants that result from the production of isothiocyanates as well as some nitriles that have been identified as possible nematicides. Previous studies have also documented the toxicity of GSLs and their hydrolysis products to nematodes [80 - 82].

The production of GLS secondary metabolites, especially GHPs, has been seen in the plants of the families Brassicaceae and Cucurbitaceae. Although crops in the Brassicaceae family are nematode hosts, they frequently prevent nematode growth and infestation [83]. The release of GHPs occurs when Brassicaceae plant tissues are applied to the soil as a result of myrosinases that are either microbial or plant-derived. High concentrations of GSL glucotropaeolin, which hydrolyzes to benzyl isothiocyanates, are found in *Brassica hirta* plants, while high concentrations of sinigrin, which hydrolyzes to allyl isothiocyanates, are found in *Brassica juncea* plants. Because of the release of both isothiocyanates into the soil caused by the burial of these plants, *Meloidogyne javanica* is less common [84]. Applying oil and dry leaf meal from *B. juncea* to tomato plants resulted in a reduction of the gall index by over 90% and a decrease in the soil populations of *Meloidogyne* species by over 60%. This was due to the release of allyl isothiocyanates into the soil [85]. Reductions in gall index and egg masses of up to 90% were observed in *Meloidogyne* spp. when *Raphanus sativus*, *Eruca sativa*, *Brassica. juncea*, *Brassica rapa*, *Brassica oleracea* var. *accephala*, *Brassica oleracea* var. *italica*, or *Brassica napus* were rotated with crops like tomato, potato, zucchini, squash, cantaloupe, or strawberry [86 - 90]. When buried, the populations of *Globodera pallida* in potato crops are decreased because *B. juncea* and *R. sativus* plants have high concentrations of sinigrin and glucoraphanin in their tissues, respectively [91].

The extracted GLS and GHPs can be applied directly to suppress the nematode invasion. In several studies, GSLs and GHPs have been applied utilising seed meals from *B. juncea* and *Brassica carinata* as nematicides primarily to prevent *Globodera* species from hatching in potato crops, and *Meloidogyne* species from hatching and forming galls in various crops and lowering *Pratylenchus* populations in apple crops [92 - 94]. Applying plant tissues such as chopped broccoli leaves [95], crude extracts [96], pellets from dry matter [97], or crushed plant tissues from *Spiraea alba* or *B. napus* and mixed with the soil [98] also demonstrated strong nematicide activity against *M. javanica*, *M. incognita*, and *Criconemoides xenoplax* in pepper, tomato, or melon. In *Abelmoschus esculentus* crops, chopped leaves from cauliflower and cabbage were similarly successful in lowering PPN populations, including root-knot, lance, spiral, and stunt nematodes, by over 80% [99]. Conversely, potential drawbacks for integrated pest management strategies involving *B. carinata* seed meal have been documented. These include potential harm to entomopathogenic nematode populations, such as *Steinernema riobrave* and *Steinernema feltiae*, which would lessen these nematodes' ability to act as biological control agents against the Colorado potato beetle (*Leptinotarsa decemlineata*) [100].

Flavonoids

All terrestrial plants include flavonoids, a broad class of secondary carbon-based compounds. Flavonoids are biostimulants and secondary metabolites that are essential to plant growth because they increase a plant's tolerance to a variety of biotic and abiotic stressors [101]. From a range of plant species, more than 10,000 distinct kinds of flavonoids have been identified. The class of phenylpropanoids known as flavonoids is produced by the shikimate and acetate pathways *via* the action of a cytosolic multienzyme complex that is attached to the endoplasmic reticulum [102]. Through the phenylpropanoid pathway and the shikimic acid pathway, 4-coumaroyl-CoA is produced during the synthesis of phenylalanine. Chalcone 2',4',6',4-tetrahydroxy synthesis occurs by the condensation process and with the aid of chalcone-synthase. After being isomerized by an isomerase (chalcone-flavanone), 2', 4',6',4-tetrahydroxy chalcone forms flavanone, which starts the flavonoid pathway and yields numerous types of flavonoids [103, 104]. Phytoalexins, flavones, flavonols, flavanones, anthocyanins, isoflavonoids, anthoxanthins, proanthocyanidins, and chalcones are among the diverse constituents of this class [104].

Among the various secondary metabolites, flavonoids, as natural polyphenolic compounds, are drawing more and more interest from scientists because they exhibit a wide range of biological actions [105]. The basic structure of flavonoids is a benzo-γ-pyrone skeleton [106]. The enzymes chalcone isomerase (CHI), chalcone synthase (CHS), and a particular form of chalcone isomerase present in legumes are involved in the production of flavonoids [107]. As a result, isoflavonoids, a characteristic component of leguminous plants, are formed. Two methods exist for secreting flavonoids into the rhizosphere: An active method that uses the ABC transporter and adenosine triphosphate (ATP) and a passive method that involves the breakdown of cortical cells and the root cap [101].

The crop production is further decreased by parasitic nematodes infecting the cysts or galls on the roots [108]. The initial contact of PPNs with flavonoids occurs when PPN finds its host and infects it. Wuyts *et al.* [109], demonstrated that kaempferol, a flavonoid, hindered and inhibited *Radopholus similis* egg hatching. Flavonoid secretion is triggered by nematode infection of plant roots [110]. Plants release cumestrol, a phytoalexin, and glyceollin, an isoflavonoid, as nematicides against *M. incognita* and *Pratylenchus penetrans*, respectively [111]. Flavonoids have a complex effect on worm behaviour; depending on their chemical structure and concentration, they can either attract or repel juvenile *M. incognita* nematodes [112]. It was demonstrated that the juveniles of *Heterodera zeae* were killed by flavonoids such as patuletin, patulitrin, quercetin, and rutin at different doses and exposure times [113]. Flavonoids that accumulate at PPN feeding sites may impact nematode reproductive potential and egg-laying capacity by restricting egg formation or skewing the ratio of males to females (*Meloidogyne* spp. and *Heterodera* spp.) [114].

In response to infection with *M. incognita*, soybeans synthesised glyceollin, a flavonoid that induces resistance, which reduced crop loss [115]. According to one study, quercetin and kaempferol were transformed by yellow-coloured cyst nematodes, *Globodera pallida* and *Globodera rostochiensis*, generating quercentagetin, a flavonoid that is specific to nematodes [116]. As a phytoalexin, cumestrol is an isoflavonoid that is produced in lima beans as a nematicide in the event of a *P. penetrans* infection. Certain flavonoids, like myricetin, kaempferol, and quercetin, have varying effects on PPNs; they can repel or attract them, stop egg hatching, cause them to go into quiescence, or even kill [106]. Flavonoids, however, have been found to have varying effects depending on the species of nematode; in fact, they have been shown to extend the life span of the model nematode, *Caenorhabditis elegans*.

Saponins

A class of plant metabolites known as saponins is frequently found in nature and possesses surfactant characteristics. Because of their natural nature and biodegradability, they are environmentally friendly [117]. Saponins are naturally occurring amphiphilic glycosides that have been reported to comprise sugars with polar glycone structure moieties separated from nonpolar aglycone structure moieties, which are also referred to as sapogenins [118]. Saponins are categorised as (i) Steroidal saponins and (ii) Triterpenoid saponins based on their aglycone counterparts [119, 120]. The plant's saponin content, which primarily consists of triterpene glycosides of medicagenic acid, has been the subject of much research. These saponins have a variety of biological functions, including the ability to biocidally affect many soil bacteria [121]. *Medicago sativa* saponins could make excellent natural nematicide formulations [122]. *In vitro* tests have shown that the saponin combinations derived from *M. sativa* tissues exhibit efficacy against three nematodes: *Globodera rostochiensis*, the root-knot worm *M. incognita*, and the virus-vector *Xiphinema index* [123].

D'Addabbo *et al.*, [124] demonstrated the bioactivity of saponins from five distinct *Medicago* species, namely, *M. heyniana*, *M. lupulina*, *M. hybrida*, *M. truncatula*, and *M. murex,* against the plant parasitic nematodes *M. incognita*, *X. index*, and *G. rostochiensis*. The study's findings show that plant biomasses and extracts rich in saponins are suppressive to nematodes that cause knots in roots. The root-knot nematodes *M. incognita* and *M. javanica* were found to be affected by the activity of saponins on juvenile (J2) motility or egg viability [124, 125]. Giannakou [126] investigated the nematicidal effect of a product that contained *Quillaja saponaria* extract against RKNs. They found that as the dose of *Q. saponaria* was increased, there was also a steady decrease in the number of juveniles emerging from egg masses of the nematode species, and they demonstrated that RKNs can be controlled by applying *Q. saponaria* extract. Previously, other economically significant phytonematode species, *X. index*, *X. americanum*, *Pratylenchus thornei*, *Meloidogyne ethiopica*, *Meloidogyne hapla*, *Helicotylenchus* sp., *Criconemoides xenoplax,* and *Tylenchorhynchus* sp., were found to be significantly affected by saponin-rich preparations from *Q. Saponaria molina* [127].

Pensec *et al.* [128], demonstrated that the use of saponins from *Gypsophila paniculata* was an effective and environmentally safe treatment against the two nematodes that spread the grapevine fanleaf virus, *i.e.*, *X. index* and *X. diversicaudatum*. Ibrahim and Srour [129] reported the management of *M. incognita* by means of saponins derived from *M. sativa*. Argentieri *et al.* [130], demonstrated the nematicidal activity against plant-parasitic nematodes possessed by saponins derived from *Medicago arabica, Medicago arborea,* and *M. sativa*.

Salicylic Acid

The significant plant hormone is salicylic acid (SA). The primary function of SA is to mediate host reactions to pathogen infections [131]. Salicylic acid is a defence-related hormone that is essential for resistance against several microbial infections [132]. Salicylic acid is the precursor to cinnamic acid and consists of a ring connected to hydroxyl and carboxyl groups [133]. Salicylic acid is present in plant roots at concentrations significantly lower than those believed to be nematistatic, despite the fact that it exhibits some nematistatic and nematicidal activity *in vitro* [134]. Rather, it appears that SA's function as a plant hormone contributes to nematode resistance. Salicylic acid signalling plays a role in genetic resistance to root-knot nematodes. The tomato root-knot nematode's defence mechanism includes salicylic acid, which is mediated by the Mi-1 gene [135] and in the regulation of different immune responses against nematodes [136]. While SA-deficient mutants exhibit greater vulnerability, pre-inoculation treatment with SA or chemical analogues thereof increases plant resistance to subsequent nematode infection [136, 137]. Resistance to the soybean cyst nematode is conferred by the overexpression of a salicylic acid methyltransferase gene in soybeans [138]. In the root of cucumbers infected with *M. javanica*, the application of SA boosted some defence chemicals, such as hydrogen peroxide and peroxidase activities [138].

Tannins

Tannins are a diverse class of polyphenolic substances with two subgroups: hydrolyzable and condensed tannins. Condensed tannins are oligomers of two or more flavan-3-ols, whereas hydrolyzable tannins have a polyol core to which galloyl groups are esterified [139]. An evaluation of the nematode-killing properties of chestnut tannin solutions on *G. rostochiensis*, a potato cyst nematode, revealed that tannin concentrations were useful in lowering the egg viability and pace of reproduction [140]. Maistrello *et al.*, [141] studied the impact of chestnut tannins on nematode (*M. javanica*) control. There was a noticeable decrease in the number of eggs and juveniles per gramme of root. Additionally, the nematode's overall population density and reproductive rates were also reduced. Similarly [142], suggested that chestnut tannins significantly reduced the viability and increased juvenile (J2) immobility. It also affected the population density and root galling index of *M. incognita*.

Phenylpropanoid

In certain plants, the shikimic acid system produces phenylpropanoids, commonly referred to as cinnamic acids, which are comparatively simple secondary metabolites [143]. This pathway is triggered by phenylalanine and tyrosine. For instance, phenylalanine ammonia-lyase (PAL) produces cinnamic acids when it removes ammonia from phenylalanine [144]. There is little research on phenylpropanoids that are involved in plants' defence against nematode infestation. Additionally, it has been demonstrated that soon after worm infection, plants activate the phenylpropanoid biosynthesis pathway's enzymes [145, 146]. Higher concentrations of phenylpropanoid compounds like chlorogenic acid in rice, tomato, and pepper roots have been reported in nematode-resistant cultivars [69, 70, 147]. In the roots of resistant bananas, phenylpropanoid enzymes, phenolic polymers, and metabolites function as chemical defences against *Pratylenchus coffeae* infection [64]. Ismail *et al.* [148], reported that phenol propanoids from *Tanacetum baltistanicum* showed nematocidal properties against *M. incognita*.

Other Compounds

Hopeaphenol has been linked to grapevine and pine tree nematode resistance. Upon infection with the pinewood worm *Bursaphelenchus xylophilus*, the stilbenoid 3-O-methyldihydropinosylvin accumulated in the bark and wood of *Pinus strobus* and had strong nematicidal activity *in vitro* [149]. It has been demonstrated that phenylphenalenone phytoalexins play a major role in banana resistance to the burrowing worm, *Radopholus similis*. These cyclic diarylheptanoids were discovered to be substantially more prevalent in the vicinity of *R. similis* infection sites in resistant banana types when banana roots were harvested 12 weeks following nematode inoculation, compared to the susceptible reference cultivar [55]. Wallis [150] investigated how host resistance to *M. incognita* is correlated with stilbenoid associations in grapevine rootstocks. They reported that RKN resistance may have been caused by the higher concentrations of particular stilbenoids in resistant individuals compared to susceptible individuals.

CONCLUSION

It is clear that PPNs are reducing global crop yields for a number of profitable crops. Therefore, more testing should be done to refine their control procedures. For the management of PPN, several facets of the interactions between plants and nematodes should be used. Numerous investigations conducted since the mid-1900s have demonstrated that plants employ a wide range of small chemical substances to protect themselves from PPN. Several inferences can be made from the research that is now accessible, despite the fact that generalisation is challenging due to the vast range of metabolites, plants, and nematode species that have been examined to date. Plants have complex, multi-layered immune systems to protect themselves from nematode attacks. The generation of small molecules with anti-worm activity, either constitutively or following nematode infection, is one aspect of plant protection against nematodes. Still, these metabolites do not completely manage the nematode problem. Developing novel strategies for PPN molecular regulation will benefit from a precise understanding of the molecular mechanisms involved in PPN interactions with plants. Despite being unprofitable in many cases right now, the use of phytochemicals in agriculture has a lot of potential.

REFERENCES

[1]   Philippot L, Raaijmakers JM, Lemanceau P, van der Putten WH. Going back to the roots: the microbial ecology of the rhizosphere. Nat Rev Microbiol 2013; 11(11): 789-99.
[http://dx.doi.org/10.1038/nrmicro3109] [PMID: 24056930]

[2]   Raaijmakers JM, Paulitz TC, Steinberg C, Alabouvette C, Moënne-Loccoz Y. The rhizosphere: A playground and battlefield for soilborne pathogens and beneficial microorganisms. Plant Soil 2009;

321(1-2): 341-61.
[http://dx.doi.org/10.1007/s11104-008-9568-6]

[3]   Kumar Y, Yadav BC. Plant-parasitic nematodes: Nature's most successful plant parasite. Int J Manag Rev 2020; 7: 379-86.

[4]   Vieira P, Gleason C. Plant-parasitic nematode effectors — insights into their diversity and new tools for their identification. Curr Opin Plant Biol 2019; 50: 37-43.
[http://dx.doi.org/10.1016/j.pbi.2019.02.007] [PMID: 30921686]

[5]   Nyaku ST, Affokpon A, Danquah A, Brentu FC. Harnessing useful rhizosphere microorganisms for nematode control. Nematology–concepts, diagnosis and control. 2017; 16: 153-182.
[http://dx.doi.org/10.5772/intechopen.69164]

[6]   Decraemer W, Hunt DJ. Structure and classification. In: Perry RN, Moens M, Eds. Plant nematology. Wallingford, UK; Cambridge, MA, USA: CABI 2006; pp. 3-32.
[http://dx.doi.org/10.1079/9781845930561.0003]

[7]   Siddiqi MR. Tylenchida: parasites of plants and insects. CABI 2000.
[http://dx.doi.org/10.1079/9780851992020.0000]

[8]   Khan MR. Nematode diseases of crops in India. In: L.P. Awasthi Ed., Recent advances in the diagnosis and management of plant diseases. 2015; 183-224.
[http://dx.doi.org/10.1007/978-81-322-2571-3_16]

[9]   Kumar D, Kumar V, Mann SS, Kamboj R. Phytonematodes problem, their occurrence, distribution, damage and management in forage and fodder crops: A review. Forage Res 2022; 47: 408-15.

[10]   Singh SK, Hodda M, Ash GJ. Plant-parasitic nematodes of potential phytosanitary importance, their main hosts and reported yield losses. Bull OEPP 2013; 43(2): 334-74.
[http://dx.doi.org/10.1111/epp.12050]

[11]   Sorribas FJ, Djian-Caporalino C, Mateille T. Nematodes. Integrated pest and disease management in greenhouse crops. 2020; 147-174.

[12]   Koenning SR, Overstreet C, Noling JW, Donald PA, Becker JO, Fortnum BA. Survey of crop losses in response to phytoparasitic nematodes in the United States for 1994. J Nematol 1999; 31(4S): 587-618.
[PMID: 19270925]

[13]   Jones MGK, Fosu-Nyarko J. Molecular biology of root lesion nematodes (*Pratylenchus* spp.) and their interaction with host plants. Ann Appl Biol 2014; 164(2): 163-81.
[http://dx.doi.org/10.1111/aab.12105]

[14]   Fosu-Nyarko J, Jones MGK. Advances in understanding the molecular mechanisms of root lesion nematode host interactions. Annu Rev Phytopathol 2016; 54(1): 253-78.
[http://dx.doi.org/10.1146/annurev-phyto-080615-100257] [PMID: 27296144]

[15]   Feyisa B. Factors associated with plant parasitic nematode (PPN) population: A review. Animal and Veterinary Sciences 2022; 10(2): 41-5.
[http://dx.doi.org/10.11648/j.avs.20221002.15]

[16]   Barker KR, Koenning SR. Developing sustainable systems for nematode management. Annu Rev Phytopathol 1998; 36(1): 165-205.
[http://dx.doi.org/10.1146/annurev.phyto.36.1.165] [PMID: 15012497]

[17]   Dong LQ, Zhang KQ. Microbial control of plant-parasitic nematodes: A five-party interaction. Plant Soil 2006; 288(1-2): 31-45.
[http://dx.doi.org/10.1007/s11104-006-9009-3]

[18]   Hassan MA, Pham TH, Shi H, Zheng J. Nematodes threats to global food security. Acta Agriculturae Scandinavica, Section B–Soil & Plant Science 2013; 63(5): 420-5.
[http://dx.doi.org/10.1080/09064710.2013.794858]

[19]   Khanna K, Kohli SK, Ohri P, Bhardwaj R. Plants-nematodes-microbes crosstalk within soil: A trade-

off among friends or foes. Microbiol Res 2021; 248: 126755.
[http://dx.doi.org/10.1016/j.micres.2021.126755] [PMID: 33845302]

[20]   Desmedt W, Mangelinckx S, Kyndt T, Vanholme B. A phytochemical perspective on plant defense against nematodes. Front Plant Sci 2020; 11: 602079.
[http://dx.doi.org/10.3389/fpls.2020.602079] [PMID: 33281858]

[21]   Sharma N, Khanna K, Jasrotia S, Kumar D, Bhardwaj R, Ohri P. Metabolites and chemical agents in the plant roots: An overview of their use in plant-parasitic nematode management. Nematology 2023; 25(3): 243-57.
[http://dx.doi.org/10.1163/15685411-bja10220]

[22]   Sikder MM, Vestergård M. Impacts of root metabolites on soil nematodes. Front Plant Sci 2020; 10: 1792.
[http://dx.doi.org/10.3389/fpls.2019.01792] [PMID: 32082349]

[23]   Mathesius U, Costa SR. Plant signals differentially affect rhizosphere nematode populations. J Exp Bot 2021; 72(10): 3496-9.
[http://dx.doi.org/10.1093/jxb/erab149] [PMID: 33948654]

[24]   Williamson VM, Gleason CA. Plant–nematode interactions. Curr Opin Plant Biol 2003; 6(4): 327-33.
[http://dx.doi.org/10.1016/S1369-5266(03)00059-1] [PMID: 12873526]

[25]   Ali MA, Azeem F, Li H, Bohlmann H. Smart parasitic nematodes use multifaceted strategies to parasitize plants. Front Plant Sci 2017; 8: 1699.
[http://dx.doi.org/10.3389/fpls.2017.01699] [PMID: 29046680]

[26]   Escobar C, Barcala M, Cabrera J, Fenoll C. Overview of root-knot nematodes and giant cells. Adv Bot Res 2015; 73: 1-32.
[http://dx.doi.org/10.1016/bs.abr.2015.01.001]

[27]   Holbein J. Deciphering the role of apoplastic root barriers in the interaction between sedentary nematodes and Arabidopsis. Doctoral dissertation, Universitäts-und Landesbibliothek Bonn 2018.

[28]   Grundler FMW, Sobczak M, Golinowski W. Formation of wall openings in root cells of *Arabidopsis thaliana* following infection by the plant-parasitic nematode *Heterodera schachtii*. Eur J Plant Pathol 1998; 104(6): 545-51.
[http://dx.doi.org/10.1023/A:1008692022279]

[29]   Moens M, Perry RN, Jones JT. Cyst nematodes-life cycle and economic importance. Cyst nematodes. Wallingford, UK: CAB International 2018; pp. 1-26.
[http://dx.doi.org/10.1079/9781786390837.0001]

[30]   Grundler FMW, Munch A, Wyss U. The parasitic behaviour of second-stage juveniles of *Meloidogyne incognita* in roots of *Arabidopsis thaliana*. Nematologica 1992; 38(1-4): 98-111.
[http://dx.doi.org/10.1163/187529292X00081]

[31]   Favery B, Quentin M, Jaubert-Possamai S, Abad P. Gall-forming root-knot nematodes hijack key plant cellular functions to induce multinucleate and hypertrophied feeding cells. J Insect Physiol 2016; 84: 60-9.
[http://dx.doi.org/10.1016/j.jinsphys.2015.07.013] [PMID: 26211599]

[32]   Jones MG, Goto DB. Root-knot nematodes and giant cells. In: Jones J, Gheysen G, Fenoll C, Eds., Genomics and molecular genetics of plant-nematode interactions. 2011; 83-100.
[http://dx.doi.org/10.1007/978-94-007-0434-3_5]

[33]   Collett RL, Rashidifard M, Marais M, Daneel M, Fourie H. Insights into the life-cycle development of *Meloidogyne enterolobii*, *M. incognita* and *M. javanica* on tomato, soybean and maize. Eur J Plant Pathol. 2024 Jan; 168(1): 137-46.
[http://dx.doi.org/10.1007/s10658-023-02741-9]

[34]   Subbotin SA, Rius JE, Castillo P. Systematics of root-knot nematodes (Nematoda: Meloidogynidae). Brill. 2021; 14: 1-853.

[35] Bartlem DG, Jones MGK, Hammes UZ. Vascularization and nutrient delivery at root-knot nematode feeding sites in host roots. J Exp Bot 2014; 65(7): 1789-98.
[http://dx.doi.org/10.1093/jxb/ert415] [PMID: 24336493]

[36] Abd-Elgawad MMM. Exploiting plant–phytonematode interactions to upgrade safe and effective nematode control. Life (Basel) 2022; 12(11): 1916.
[http://dx.doi.org/10.3390/life12111916] [PMID: 36431051]

[37] Gillet FX, Bournaud C, de Souza Júnior JDA, Fatima Grossi-de-Sa M. Plant-parasitic nematodes: towards understanding molecular players in stress responses. Ann Bot (Lond) 2017; 119(5): mcw260.
[http://dx.doi.org/10.1093/aob/mcw260] [PMID: 28087659]

[38] Khan M, Khan AU. Plant parasitic nematodes effectors and their crosstalk with defense response of host plants: A battle underground. Rhizosphere 2021; 17: 100288.
[http://dx.doi.org/10.1016/j.rhisph.2020.100288]

[39] Sato K, Kadota Y, Shirasu K. Plant immune responses to parasitic nematodes. Front Plant Sci 2019; 10: 1165.
[http://dx.doi.org/10.3389/fpls.2019.01165] [PMID: 31616453]

[40] Siddique S, Coomer A, Baum T, Williamson VM. Recognition and response in plant–nematode interactions. Annu Rev Phytopathol 2022; 60(1): 143-62.
[http://dx.doi.org/10.1146/annurev-phyto-020620-102355] [PMID: 35436424]

[41a] Viant MR, Kurland IJ, Jones MR, Dunn WB. How close are we to complete annotation of metabolomes? Curr Opin Chem Biol 2017; 36: 64-9.
[http://dx.doi.org/10.1016/j.cbpa.2017.01.001] [PMID: 28113135]

[41b] Singh S. Natural product chemistry in agriculture. In: Kumar V, Tsatsaragkou K, Asim N, Eds. Green chemistry in agriculture and food production. Taylor & francis, CRC Press 2023; pp. 1: 28.
[http://dx.doi.org/10.1201/9780429289538]

[42] Renčo M, Sasanelli N, Maistrello L. Plants as natural sources of nematicides In: Davis LM Ed, Nematodes: Comparative genomics, disease management and ecological importance Nematicides. New York: NOVA Science publisher 2014; pp. 115-41.

[43] Sikder MM, Vestergård M, Kyndt T, Kudjordjie EN, Nicolaisen M. Phytohormones selectively affect plant parasitic nematodes associated with *Arabidopsis* roots. New Phytol 2021; 232(3): 1272-85.
[http://dx.doi.org/10.1111/nph.17549] [PMID: 34115415]

[44] Boudet AM. Evolution and current status of research in phenolic compounds. Phytochemistry 2007; 68(22-24): 2722-35.
[http://dx.doi.org/10.1016/j.phytochem.2007.06.012] [PMID: 17643453]

[45] Lattanzio V, Kroon PA, Quideau S, Treutter D. Plant phenolics—secondary metabolites with diverse functions. Rec Adv Polyphen Res 2009; 1: 1-35.

[46] Stiller A, Garrison K, Gurdyumov K, *et al.* From fighting critters to saving lives: polyphenols in plant defense and human health. Int J Mol Sci 2021; 22(16): 8995.
[http://dx.doi.org/10.3390/ijms22168995] [PMID: 34445697]

[47] Marchiosi R, dos Santos WD, Constantin RP, *et al.* Biosynthesis and metabolic actions of simple phenolic acids in plants. Phytochem Rev 2020; 19(4): 865-906.
[http://dx.doi.org/10.1007/s11101-020-09689-2]

[48] Lattanzio V, Lattanzio VM, Cardinali A. Role of phenolics in the resistance mechanisms of plants against fungal pathogens and insects. Adv Photochem 2006; 661: 23-67.

[49] Aissani N, Balti R, Sebai H. Potent nematicidal activity of phenolic derivatives on *Meloidogyne incognita*. J Helminthol 2018; 92(6): 668-73.
[http://dx.doi.org/10.1017/S0022149X17000918] [PMID: 29017629]

[50] Ohri P, Pannu SK. Effect of phenolic compounds on nematodes- A review. J Appl Nat Sci 2010; 2(2):

344-50.
[http://dx.doi.org/10.31018/jans.v2i2.144]

[51]   Albuquerque EVS, Carneiro RMDG, Costa PM, *et al.* Resistance to *Meloidogyne incognita* expresses a hypersensitive-like response in *Coffea arabica.* Eur J Plant Pathol 2010; 127(3): 365-73.
[http://dx.doi.org/10.1007/s10658-010-9603-3]

[52]   Bhattacharyya R, Pandey SC, Chandra S, *et al.* Fertilization effects on yield sustainability and soil properties under irrigated wheat–soybean rotation of an Indian Himalayan upper valley. Nutr Cycl Agroecosyst 2010; 86(2): 255-68.
[http://dx.doi.org/10.1007/s10705-009-9290-7]

[53]   Ahmad G, Khan A, Khan AA, Ali A, Mohhamad HI. Biological control: A novel strategy for the control of the plant parasitic nematodes. Antonie van Leeuwenhoek 2021; 114(7): 885-912.
[http://dx.doi.org/10.1007/s10482-021-01577-9] [PMID: 33893903]

[54]   Dhakshinamoorthy S, Mariama K, Elsen A, De Waele D. Phenols and lignin are involved in the defence response of banana (*Musa*) plants to *Radopholus similis* infection. Nematology 2014; 16(5): 565-76.
[http://dx.doi.org/10.1163/15685411-00002788]

[55]   Hölscher D, Dhakshinamoorthy S, Alexandrov T, *et al.* Phenalenone-type phytoalexins mediate resistance of banana plants (*Musa* spp.) to the burrowing nematode *Radopholus similis.* Proc Natl Acad Sci USA 2014; 111(1): 105-10.
[http://dx.doi.org/10.1073/pnas.1314168110] [PMID: 24324151]

[56]   Yates P, Janiol J, Li C, Song BH. Nematocidal potential of phenolic acids: a phytochemical seed-coating approach to soybean cyst nematode management. Plants 2024; 13(2): 319.
[http://dx.doi.org/10.3390/plants13020319] [PMID: 38276776]

[57]   Caboni P, Aissani N, Demurtas M, Ntalli N, Onnis V. Nematicidal activity of acetophenones and chalcones against *Meloidogyne incognita* and structure–activity considerations. Pest Manag Sci 2016; 72(1): 125-30.
[http://dx.doi.org/10.1002/ps.3978] [PMID: 25641877]

[58]   Oliveira DF, Costa VA, Terra WC, Campos VP, Paula PM, Martins SJ. Impact of phenolic compounds on *Meloidogyne incognita in vitro* and in tomato plants. Exp Parasitol 2019; 199: 17-23.
[http://dx.doi.org/10.1016/j.exppara.2019.02.009] [PMID: 30790574]

[59]   Gupta RL, Prasad D, Thukral R. Quantitative structure activity relationship study for the fungicidal and nematicidal activity of phenols. Pestic Res J 2005; 17: 15-20.

[60]   Balaji A, Kannan S. Impact of different phenolic compounds on hatchability of *Meloidogyne incognita.* Geobios 1988; 15: 143-4.

[61]   Mahajan RA, Singh PR, Bajaj KL. Nematicidal activity of some phenolic compounds against *Meloidogyne incognita.* Revue de Nematologie 1985; 8: 161-4.

[62]   Acedo JR, Rohde RA. Histochemical root pathology of *Brassica oleracea capitata* L. infected by *Pratylenchus penetrans* (Cobb) Filipjev and Schuurmans Stekhoyen (Nematoda: Tylenchidae). J Nematol 1971; 3(1): 62-8.
[PMID: 19322342]

[63]   Fabiyi OA, Atolani O, Olatunji GA. Toxicity effect of *Eucalyptus globulus* on *Pratylenchus* spp. of *Zea mays.* Sarhad J Agric 2020; 36(4): 1244-53.
[http://dx.doi.org/10.17582/journal.sja/2020.36.4.1244.1253]

[64]   Vaganan MM, Ravi I, Nandakumar A, Sarumathi S, Sundararaju P, Mustaffa MM. Phenylpropanoid enzymes, phenolic polymers and metabolites as chemical defenses to infection of *Pratylenchus coffeae* in roots of resistant and susceptible bananas (*Musa* spp.). Indian J Exp Biol 2014; 52(3): 252-60.
[PMID: 24669668]

[65]   Kundu A, Vadassery J. Chlorogenic acid-mediated chemical defence of plants against insect

herbivores. Plant Biol 2019; 21(2): 185-9.
[http://dx.doi.org/10.1111/plb.12947] [PMID: 30521134]

[66] Tunnicliffe JM, Cowan T, Shearer J. Chlorogenic acid in whole body and tissue-specific glucose regulation. In: Preedy VR Ed., Coffee in health and disease prevention. 2015; 1: 777-785.
[http://dx.doi.org/10.1016/B978-0-12-409517-5.00086-3]

[67] Hajam YA. Diksha, Kumar R., Lone R. Plant phenolics and their versatile promising role in the management of nematode stress. In: Lone R, Khan S, Al-Sadi AM, Eds. Plant Phenolics in Biotic Stress Management. Springer 2024; pp. 389-416.
[http://dx.doi.org/10.1007/978-981-99-3334-1_16]

[68] Meher HC, Gajbhiye VT, Singh G, Chawla G. Altered metabolomic profile of selected metabolites and improved resistance of *Cicer arietinum* (L.) against *Meloidogyne incognita* (Kofoid & White) Chitwood following seed soaking with salicylic acid, benzothiadiazole or nicotinic acid. Acta Physiol Plant 2015; 37(7): 140.
[http://dx.doi.org/10.1007/s11738-015-1888-6]

[69] Pegard A, Brizzard G, Fazari A, Soucaze O, Abad P, Djian-Caporalino C. Histological characterization of resistance to different root-knot nematode species related to phenolics accumulation in *Capsicum annuum*. Phytopathology 2005; 95(2): 158-65.
[http://dx.doi.org/10.1094/PHYTO-95-0158] [PMID: 18943985]

[70] Gill JR, Harbornez JB, Plowright RA, Grayer RJ, Rahman ML. The induction of phenolic compounds in rice after infection by the stem nematode *Ditylenchus angustus*. Nematologica 1996; 42(5): 564-78.
[http://dx.doi.org/10.1163/004625996X00063]

[71] Baldacci-Cresp F, Behr M, Kohler A, *et al*. Molecular changes concomitant with vascular system development in mature galls induced by root-knot nematodes in the model tree host *Populus tremula× P. alba*. Int J Mol Sci 2020; 21(2): 406.
[http://dx.doi.org/10.3390/ijms21020406] [PMID: 31936440]

[72] D'Addabbo T, Carbonara T, Argentieri MP, *et al*. Nematicidal potential of *Artemisia annua* and its main metabolites. Eur J Plant Pathol 2013; 137(2): 295-304.
[http://dx.doi.org/10.1007/s10658-013-0240-5]

[73] Mnviajan R, Singh P, Krishan L. Nematicidal activity of some phenolic compounds. Revue Nématol 1985; 8: 161-4.

[74] Afifah EN, Murti RH, Nuringtyas TR. Metabolomics approach for the analysis of resistance of four tomato genotypes (*Solanum lycopersicum* L.) to root-knot nematodes (*Meloidogyne incognita*). Open Life Sci 2019; 14(1): 141-9.
[http://dx.doi.org/10.1515/biol-2019-0016] [PMID: 33817146]

[75] Mitreiter S, Gigolashvili T. Regulation of glucosinolate biosynthesis. J Exp Bot 2021; 72(1): 70-91.
[http://dx.doi.org/10.1093/jxb/eraa479] [PMID: 33313802]

[76] Francisco M, Joseph B, Caligagan H, *et al*. The defense metabolite, allyl glucosinolate, modulates *Arabidopsis thaliana* biomass dependent upon the endogenous glucosinolate pathway. Front Plant Sci 2016; 7: 774.
[http://dx.doi.org/10.3389/fpls.2016.00774] [PMID: 27313596]

[77] Francisco M, Joseph B, Caligagan H, *et al*. Genome wide association mapping in *Arabidopsis thaliana* identifies novel genes involved in linking allyl glucosinolate to altered biomass and defense. Front Plant Sci 2016; 7: 1010.
[http://dx.doi.org/10.3389/fpls.2016.01010] [PMID: 27462337]

[78] Avato P, D'Addabbo T, Leonetti P, Argentieri MP. Nematicidal potential of brassicaceae. Phytochem Rev 2013; 12(4): 791-802.
[http://dx.doi.org/10.1007/s11101-013-9303-7]

[79] Eugui D, Escobar C, Velasco P, Poveda J. Glucosinolates as an effective tool in plant-parasitic

nematodes control: Exploiting natural plant defenses. Appl Soil Ecol 2022; 176: 104497.
[http://dx.doi.org/10.1016/j.apsoil.2022.104497]

[80] Buskov S, Serra B, Rosa E, Sørensen H, Sørensen JC. Effects of intact glucosinolates and products produced from glucosinolates in myrosinase-catalyzed hydrolysis on the potato cyst nematode (*Globodera rostochiensis* Cv. Woll). J Agric Food Chem 2002; 50(4): 690-5.
[http://dx.doi.org/10.1021/jf010470s] [PMID: 11829629]

[81] Lazzeri L, Tacconi R, Palmieri S. *In vitro* activity of some glucosinolates and their reaction products toward a population of the nematode *Heterodera schachtii*. J Agric Food Chem 1993; 41(5): 825-9.
[http://dx.doi.org/10.1021/jf00029a028]

[82] Zasada IA, Ferris H. Sensitivity of *Meloidogyne javanica* and *Tylenchulus semipenetrans* to isothiocyanates in laboratory assays. Phytopathology 2003; 93(6): 747-50.
[http://dx.doi.org/10.1094/PHYTO.2003.93.6.747] [PMID: 18943062]

[83] Kirkegaard J, McLeod R, Steel C. Invasion, development, growth and egg laying by *Meloidogyne javanica* in Brassicaceae crops. Nematology 2001; 3(5): 463-72.
[http://dx.doi.org/10.1163/156854101753250791]

[84] Zasada IA, Ferris H. Nematode suppression with brassicaceous amendments: Application based upon glucosinolate profiles. Soil Biol Biochem 2004; 36(7): 1017-24.
[http://dx.doi.org/10.1016/j.soilbio.2003.12.014]

[85] Oliveira RDL, Dhingra OD, Lima AO, *et al.* Glucosinolate content and nematicidal activity of Brazilian wild mustard tissues against *Meloidogyne incognita* in tomato. Plant Soil 2011; 341(1-2): 155-64.
[http://dx.doi.org/10.1007/s11104-010-0631-8]

[86] Al-Rehiayani S, Hafez S. Host status and green manure effect of selected crops on *Meloidogyne chitwoodi* race 2 and *Pratylenchus neglectus*. Nematropica 1998; 28: 213-30.

[87] Aydınlı G, Mennan S. Yield and resistance of tomato rootstocks to *Meloidogyne arenaria* in a greenhouse. Pesqui Agropecu Bras 2022; 57: e02418.
[http://dx.doi.org/10.1590/s1678-3921.pab2022.v57.02418]

[88] López-Pérez JA, Escuer M, Díez-Rojo MA, Robertson L, Buena AP, López-Cepero J, Bello A. Host range of *Meloidogyne arenaria* (Neal, 1889) Chitwood, 1949 (Nematoda: Meloidogynidae) in Spain [Hospedadores de *Meloidogyne arenaria* (Neal, 1889) Chitwood, 1949) (Nematoda: Meloidogynidae) en España]. Nematropica. 2011 Jun 1;41(2):130–40.

[89] Mojtahedi H, Santo GS, Wilson JH, Hang AN. Managing *Meloidogyne chitwoodi* on potato with rapeseed as green manure. Plant Dis 1993; 77(1): 42-6.
[http://dx.doi.org/10.1094/PD-77-0042]

[90] Monfort WS, Kirkpatrick TL, Rothrock CS, Mauromoustakos A. Potential for site-specific management of *Meloidogyne incognita* in cotton using soil textural zones. J Nematol 2007; 39(1): 1-8.
[PMID: 19259468]

[91] Ngala BM, Woods SR, Back MA. *In vitro* assessment of the effects of *Brassica juncea* and *Raphanus sativus* leaf and root extracts on the viability of *Globodera pallida* encysted eggs. Nematology 2015; 17(5): 543-56.
[http://dx.doi.org/10.1163/15685411-00002888]

[92] Dandurand LM, Morra MJ, Zasada IA, Phillips WS, Popova I, Harder C. Control of *Globodera* spp. using *Brassica juncea* seed meal and seed meal extract. J Nematol .2017; 49: 437-45.
[PMID: 29353933]

[93] Handiseni M, Cromwell W, Zidek M, Zhou XG, Jo YK. Use of brassicaceous seed meal extracts for managing root-knot nematode in bermudagrass. Nematropica 2017; 47: 55-62.

[94] Mocali S, Landi S, Curto G, *et al.* Resilience of soil microbial and nematode communities after biofumigant treatment with defatted seed meals. Ind Crops Prod 2015; 75: 79-90.

[http://dx.doi.org/10.1016/j.indcrop.2015.04.031]

[95] Ploeg A, Stapleton J. Glasshouse studies on the effects of time, temperature and amendment of soil with broccoli plant residues on the infestation of melon plants by *Meloidogyne incognita* and *M. javanica*. Nematology 2001; 3(8): 855-61.
[http://dx.doi.org/10.1163/156854101753625353]

[96] Mashela PW, Dube ZP, Pofu KM. Managing the phytotoxicity and inconsistent nematode suppression in soil amended with phytonematicides. In: Meghvansi MK, Varma A Eds., Organic amendments and soil suppressiveness in plant disease management. 2015; 147-173.
[http://dx.doi.org/10.1007/978-3-319-23075-7_7]

[97] Guerrero-Díaz MM, Lacasa-Martínez CM, Hernández-Piñera A, Martínez-Alarcón V, Lacasa-Plasencia A. Evaluation of repeated biodisinfestation using *Brassica carinata* pellets to control *Meloidogyne incognita* in protected pepper crops. Span J Agric Res 2013; 11(2): 485-93.
[http://dx.doi.org/10.5424/sjar/2013112-3275]

[98] Kruger DH, Fourie JC, Malan AP. The effect of cover crops and their management on plant-parasitic nematodes in vineyards. S Afr J Enol Vitic 2015; 36: 195-209.

[99] Behera SK, Sahu A, Das N, Kumari A. Effect of bio-fumigation on nematode population and nutrient status of soil in okra. J Entomol Zool Stud 2020; 8: 394-7.

[100] Henderson DR, Riga E, Ramirez RA, Wilson J, Snyder WE. Mustard biofumigation disrupts biological control by *Steinernema* spp. nematodes in the soil. Biol Control 2009; 48(3): 316-22.
[http://dx.doi.org/10.1016/j.biocontrol.2008.12.004]

[101] Shah A, Smith DL. Flavonoids in agriculture: Chemistry and roles in, biotic and abiotic stress responses, and microbial associations. Agronomy (Basel) 2020; 10(8): 1209.
[http://dx.doi.org/10.3390/agronomy10081209]

[102] Petrussa E, Braidot E, Zancani M, *et al.* Plant flavonoids-biosynthesis, transport and involvement in stress responses. Int J Mol Sci 2013; 14(7): 14950-73.
[http://dx.doi.org/10.3390/ijms140714950] [PMID: 23867610]

[103] Cesco S, Mimmo T, Tonon G, *et al.* Plant-borne flavonoids released into the rhizosphere: impact on soil bio-activities related to plant nutrition. A review. Biol Fertil Soils 2012; 48(2): 123-49.
[http://dx.doi.org/10.1007/s00374-011-0653-2]

[104] Liga S, Paul C, Péter F. Flavonoids: Overview of biosynthesis, biological activity, and current extraction techniques. Plants 2023; 12(14): 2732.
[http://dx.doi.org/10.3390/plants12142732] [PMID: 37514347]

[105] Chin S, Behm CA, Mathesius U. Functions of flavonoids in plant–nematode interactions. Plants 2018; 7(4): 85.
[http://dx.doi.org/10.3390/plants7040085] [PMID: 30326617]

[106] Kumar GA, Kumar S, Bhardwaj R, *et al.* Recent advancements in multifaceted roles of flavonoids in plant–rhizomicrobiome interactions. Front Plant Sci 2024; 14: 1297706.
[http://dx.doi.org/10.3389/fpls.2023.1297706] [PMID: 38250451]

[107] Liu CW, Murray JD. The role of flavonoids in nodulation host-range specificity: An update. Plants 2016; 5(3): 33.
[http://dx.doi.org/10.3390/plants5030033] [PMID: 27529286]

[108] Mandal HR, Katel S, Subedi S, Shrestha J. Plant Parasitic Nematodes and their management in crop production: A review. J Agric Nat Resour 2021; 4(2): 327-38.
[http://dx.doi.org/10.3126/janr.v4i2.33950]

[109] Wuyts N, Lognay G, Swennen R, De Waele D. Nematode infection and reproduction in transgenic and mutant *Arabidopsis* and tobacco with an altered phenylpropanoid metabolism. J Exp Bot 2006; 57(11): 2825-35.
[http://dx.doi.org/10.1093/jxb/erl044] [PMID: 16831845]

[110] Bano S, Iqbal EY, Lubna , Zik-ur-Rehman S, Fayyaz S, Faizi S. Nematicidal activity of flavonoids with structure activity relationship (SAR) studies against root knot nematode *Meloidogyne incognita*. Eur J Plant Pathol 2020; 157(2): 299-309.
[http://dx.doi.org/10.1007/s10658-020-01988-w]

[111] San Chin SL. The role of flavonoids in the interaction between the plant, *Medicago truncatula* and the nematode, *Meloidogyne javanica* (PhD dissertation, The Australian National University 2019.

[112] Kirwa HK. Behavioral response of root-knot nematode, *Meloidogyne incognita*, to tomato, Solanum lycopersicon, root exudates. M. Sc. Thesis submitted to Jomo Kenyatta Universty of Agriculture And Technology 2020.

[113] Faizi S, Fayyaz S, Bano S, *et al.* Isolation of nematicidal compounds from *Tagetes patula* L. yellow flowers: structure-activity relationship studies against cyst nematode *Heterodera zeae* infective stage larvae. J Agric Food Chem 2011; 59(17): 9080-93.
[http://dx.doi.org/10.1021/jf201611b] [PMID: 21780738]

[114] Grundler F, Schnibbe L, Wyss U. *In vitro* studies on the behaviour of second-stage juveniles of *Heterodera schachtii* (Nematoda: Heteroderidae) in response to host plant root exudates. Parasitology 1991; 103(1): 149-55.
[http://dx.doi.org/10.1017/S0031182000059394]

[115] Kaplan DT, Keen NT, Thomason IJ. Studies on the mode of action of glyceollin in soybean incompatibility to the root knot nematode, *Meloidogyne incognita*. Physiol Plant Pathol 1980; 16(3): 319-25.
[http://dx.doi.org/10.1016/S0048-4059(80)80003-8]

[116] Evangelos G, Smith L. Flavonoids in potato cyst nematodes. Fundam Appl Nematol 1993; 16: 103-6.

[117] Rai S, Acharya-Siwakoti E, Kafle A, Devkota HP, Bhattarai A. Plant-derived saponins: A review of their surfactant properties and applications. Sci 2021; 3(4): 44.
[http://dx.doi.org/10.3390/sci3040044]

[118] Sparg SG, Light ME, van Staden J. Biological activities and distribution of plant saponins. J Ethnopharmacol 2004; 94(2-3): 219-43.
[http://dx.doi.org/10.1016/j.jep.2004.05.016] [PMID: 15325725]

[119] Moghimipour E, Handali S. Saponin: properties, methods of evaluation and applications. Annu Res Rev Biol 2015; 5(3): 207-20.
[http://dx.doi.org/10.9734/ARRB/2015/11674]

[120] Nakayasu M, Yamazaki S, Aoki Y, Yazaki K, Sugiyama A. Triterpenoid and steroidal saponins differentially influence soil bacterial genera. Plants 2021; 10(10): 2189.
[http://dx.doi.org/10.3390/plants10102189] [PMID: 34685998]

[121] Sharma P, Tyagi A, Bhansali P, *et al.* Saponins: Extraction, bio-medicinal properties and way forward to anti-viral representatives. Food Chem Toxicol 2021; 150: 112075.
[http://dx.doi.org/10.1016/j.fct.2021.112075] [PMID: 33617964]

[122] D'Addabbo T, Carbonara T, Leonetti P, Radicci V, Tava A, Avato P. Control of plant parasitic nematodes with active saponins and biomass from *Medicago sativa*. Phytochem Rev 2011; 10(4): 503-19.
[http://dx.doi.org/10.1007/s11101-010-9180-2]

[123] D'Addabbo T, Argentieri MP, Żuchowski J, *et al.* Activity of saponins from *Medicago* species against phytoparasitic nematodes. Plants 2020; 9(4): 443.
[http://dx.doi.org/10.3390/plants9040443] [PMID: 32252361]

[124] Chitwood DJ. Phytochemical based strategies for nematode control. Annu Rev Phytopathol 2002; 40(1): 221-49.
[http://dx.doi.org/10.1146/annurev.phyto.40.032602.130045] [PMID: 12147760]

[125]   Omar SA, Abdel-Massih MI, Mohamed BE. Use of saponin to control the root-knot nematode, *Meloidogyne javanica* in tomato plants. Bulletin of Faculty of Agriculture. University of Cairo 1994; 45: 933-40.

[126]   Giannakou IO. Efficacy of a formulated product containing *Quillaja saponaria* plant extracts for the control of root-knot nematodes. Eur J Plant Pathol 2011; 130(4): 587-96.
[http://dx.doi.org/10.1007/s10658-011-9780-8]

[127]   Magunacelaya JC, Martín RS. Control of plant-parasitic nematodes with extracts of *Quillaja saponaria.* Nematology 2005; 7(4): 577-85.
[http://dx.doi.org/10.1163/156854105774384732]

[128]   Pensec F, Marmonier A, Marchal A, *et al. Gypsophila paniculata* root saponins as an environmentally safe treatment against two nematodes, natural vectors of grapevine fanleaf degeneration. Aust J Grape Wine Res 2013; 19: 439-45.

[129]   Ibrahim MA, Srour HA. Saponins suppress nematode cholesterol biosynthesis and inhibit root knot nematode development in tomato seedlings. Nat Prod Chem Res 2013; 2: 1-6.

[130]   Argentieri MP, D'Addabbo T, Tava A, Agostinelli A, Jurzysta M, Avato P. Evaluation of nematicidal properties of saponins from *Medicago* spp. Eur J Plant Pathol 2008; 120(2): 189-97.
[http://dx.doi.org/10.1007/s10658-007-9207-8]

[131]   Lefevere H, Bauters L, Gheysen G. Salicylic acid biosynthesis in plants. Front Plant Sci 2020; 11: 338.
[http://dx.doi.org/10.3389/fpls.2020.00338] [PMID: 32362901]

[132]   Vlot AC, Dempsey DMA, Klessig DF. Salicylic Acid, a multifaceted hormone to combat disease. Annu Rev Phytopathol 2009; 47(1): 177-206.
[http://dx.doi.org/10.1146/annurev.phyto.050908.135202] [PMID: 19400653]

[133]   Hassoon AS, Abduljabbar IA. Review on the role of salicylic acid in plants. sustain. crop prod. 2019; 23: 61-64.

[134]   Wuyts N, Swennen R, De Waele D. Effects of plant phenylpropanoid pathway products and selected terpenoids and alkaloids on the behaviour of the plant-parasitic nematodes *Radopholus similis, Pratylenchus penetrans* and *Meloidogyne incognita.* Nematology 2006; 8(1): 89-101.
[http://dx.doi.org/10.1163/156854106776179953]

[135]   Branch C, Hwang CF, Navarre DA, Williamson VM. Salicylic acid is part of the Mi-1-mediated defense response to root-knot nematode in tomato. Mol Plant Microbe Interact 2004; 17(4): 351-6.
[http://dx.doi.org/10.1094/MPMI.2004.17.4.351] [PMID: 15077667]

[136]   Martínez-Medina A, Fernandez I, Lok GB, Pozo MJ, Pieterse CMJ, Van Wees SCM. Shifting from priming of salicylic acid- to jasmonic acid-regulated defences by *Trichoderma* protects tomato against the root knot nematode *Meloidogyne incognita.* New Phytol 2017; 213(3): 1363-77.
[http://dx.doi.org/10.1111/nph.14251] [PMID: 27801946]

[137]   Uehara T, Sugiyama S, Matsuura H, Arie T, Masuta C. Resistant and susceptible responses in tomato to cyst nematode are differentially regulated by salicylic acid. Plant Cell Physiol 2010; 51(9): 1524-36.
[http://dx.doi.org/10.1093/pcp/pcq109] [PMID: 20660227]

[138]   Lin J, Mazarei M, Zhao N, *et al.* Overexpression of a soybean salicylic acid methyltransferase gene confers resistance to soybean cyst nematode. Plant Biotechnol J 2013; 11(9): 1135-45.
[http://dx.doi.org/10.1111/pbi.12108] [PMID: 24034273]

[139]   Mora J, Pott DM, Osorio S, Vallarino JG. Regulation of plant tannin synthesis in crop species. Front Genet 2022; 13: 870976.
[http://dx.doi.org/10.3389/fgene.2022.870976] [PMID: 35586570]

[140]   Renčo M, Sasanelli N, Papajová I, Maistrello L. Nematicidal effect of chestnut tannin solutions on the potato cyst nematode *Globodera rostochiensis* (Woll.) Barhens. Helminthologia 2012; 49(2): 108-14.
[http://dx.doi.org/10.2478/s11687-012-0022-1]

[141] Maistrello L, Vaccari G, Sasanelli N. Effect of chestnut tannins on the root-knot nematode *Meloidogyne javanica.* Helminthologia 2010; 47(1): 48-57.
[http://dx.doi.org/10.2478/s11687-010-0008-9]

[142] d'Errico G, Woo SL, Lombardi N, Manganiello G, Roversi PF. Activity of chestnut tannins against the southern root-knot nematode *Meloidogyne incognita.* Redia (Firenze) 2018; 101: 53-9.
[http://dx.doi.org/10.19263/REDIA-101.18.08]

[143] Hamamouch N, Adil E. The role of the shikimate and the phenylpropanoid pathways in root-knot nematode infection. Prog Bot 2019; 81: 307-21.
[http://dx.doi.org/10.1007/124_2019_31]

[144] Seigler DS, Seigler DS. Phenylpropanoids. Plant secondary metabolism. 1998; 106-129.

[145] Baldridge GD, O'Neill NR, Samac DA. Alfalfa (*Medicago sativa* L.) resistance to the root-lesion nematode, *Pratylenchus penetrans*: defense-response gene mRNA and isoflavonoid phytoalexin levels in roots. Plant Mol Biol 1998; 38(6): 999-1010.
[http://dx.doi.org/10.1023/A:1006182908528] [PMID: 9869406]

[146] Edens RM, Anand SC, Bolla RI. Enzymes of the phenylpropanoid pathway in soybean infected with *Meloidogyne incognita or Heterodera glycines.* J Nematol 1995; 27(3): 292-303.
[PMID: 19277292]

[147] Hung C, Rohde RA. Phenol accumulation related to resistance in tomato to infection by root-knot and lesion nematodes. J Nematol 1973; 5(4): 253-8.
[PMID: 19319346]

[148] Ismail M, Javed S, Kazim M, *et al.* Phenylpropanoids from *Tanacetum baltistanicum* with nematocidal and insecticidal activities. Chem Nat Compd 2022; 58(4): 637-43.
[http://dx.doi.org/10.1007/s10600-022-03759-x]

[149] Hanawa F, Yamada T, Nakashima T. Phytoalexins from *Pinus strobus.* bark infected with pinewood nematode, *Bursaphelenchus xylophilus.* Phytochemistry 2001; 57(2): 223-8.
[http://dx.doi.org/10.1016/S0031-9422(00)00514-8] [PMID: 11382237]

[150] Wallis CM. Grapevine (*Vitis* spp.) rootstock stilbenoid associations with host resistance to and induction by root knot nematodes, *Meloidogyne incognita.* BMC Res Notes 2020; 13(1): 360.
[http://dx.doi.org/10.1186/s13104-020-05201-3] [PMID: 32727572]

CHAPTER 8

Current Limitations and Challenges in Utilizing Phytochemicals as Plant Defense Mechanism Against Nematodes

Vaseem Raja[1,*], Mudasir Ahmad Mir[2] and Aashaq Hussain Bhat[3]

[1] *University Centre for Research and Development, Chandigarh University, Gharuan, Mohali 140413, Punjab, India*

[2] *University Institute of Biotechnology, Chandigarh University, Gharuan, Mohali 140413, Punjab, India*

[3] *Department of Biosciences, University Centre for Research and Development (UCRD), Chandigarh University, Gharuan, Mohali 140413, Punjab, India*

Abstract: Understanding the dynamic interplay between phytochemicals and nematodes is vital for advancing integrated pest management strategies. Phytochemicals, the naturally occurring compounds in plants, have garnered significant attention for their potential role in defense against plant-parasitic nematodes. These bioactive compounds can deter nematodes through various mechanisms, including toxicity, repellence, and interference with nematode development. Despite promising laboratory results, the practical application of phytochemicals in agriculture faces several limitations and challenges. One major challenge is variability in phytochemicals' production among plant species and even within different parts of the same plant, influenced by environmental factors and genetic variability. Furthermore, the complex interactions between phytochemicals and the soil microbiome can impact their efficacy and stability. Another significant hurdle is the potential for nematodes to develop resistance over time, reducing the long-term effectiveness of these compounds. Additionally, the extraction, formulation, and application methods of phytochemicals must be optimized to ensure they are cost-effective and environmentally sustainable. Addressing these challenges requires multidisciplinary approaches, integrating plant breeding, molecular biology, soil science, and agronomy to develop reliable and robust phytochemical-based strategies for nematode management.

Keywords: Alkaloids, Diarylheptanoids, Flavonoids, Hydroxycinnamic acids, Nematodes, Phytochemicals, Phenolics, Plant defense, Plant parasitic nematodes, Stilbenoids, Tannins.

* **Corresponding author Vaseem Raja:** University Centre for Research and Development, Chandigarh University, Gharuan, Mohali 140413, Punjab, India; Tel: +917006968647;
E-mails: wrajamp2009@gmail.com, Vaseem.e14141@cumail.in

Shivam Jasrotia & Ajay Kumar (Eds.)

INTRODUCTION

Phytopathogens, which are organisms that cause diseases in plants, typically attack during the growth phase, leading to significant disruptions in cellular metabolism and nutrient absorption. These organisms are common in cultivated and postharvest cereals, vegetables, and fruits. Plant-parasitic nematodes (PPNs) rank are amongsome of the most significant agricultural pests, as they reduce crop yields by altering nutrient uptake and provoking secondary infections as well as acting as vectors for viruses and distortion of water transport [1, 2]. While nematode parasitism rarely results in plant death, the impact of PPNs on agriculture is considerable. Though quantifying their exact influence is challenging, estimates suggest that PPNs contribute to a global yield reduction of 10-25% [3]. Around four thousand different species of nematodes that are parasitic to plants were recorded, with the majority feeding on roots and others on above-ground plant components [4]. Despite the high degree of diversity among them, only a few genera within the sedentary PPNs—cyst and root-knot nematodes—cause the great bulk of losses [5]. These nematodes are very hard to control, and management needs a combinative approach comprising some chemical treatments, cultural practices, biocontrol, and planting highly resistant planters, which, however practicable, is additionally crucial [6]. Plant nematologists have devoted significant resources to studying plant defense mechanisms against PPNs to develop innovative control methods. One such strategy involves the synthesis of metabolites with anti-nematode properties [7].

Despite their effectiveness and accessibility, pesticides are commonly used to combat PPNs. However, they have several drawbacks, including the development of resistance and their classification as toxic substances. This toxicity affects not only bacteria, fungi, viruses, protozoa, and nematodes but also poses risks to all living organisms including humans, animals, and the environment [8]. The pesticide usage can result in both acute and chronic toxicity, persisting in the environment and leading to soil and water pollution. Furthermore, their incorporation into the food chain can lead to bioaccumulation and biomagnification. These substances have been linked to various toxic mechanisms, including acting as endocrine disruptors and generating oxidative stress within cells [9]. They have acquired this developmental ability due to the evolution of more advanced secondary metabolism in plants, which allows photosynthetic organisms to synthesize a plethora of metabolites. Plants are thought to produce 200,000 secondary metabolites [10].

OVERVIEW OF KNOWN ANTI-NEMATODE COMPOUNDS (ANPS)

Anti-nematode compounds (ANPs) are categorized into various classes, including saponins, glucosinolates, alkaloids, terpenoids, organosulfur compounds, and phenolic compounds (Fig. **1**). Although some chemicals may fall into more than one category, this taxonomy offers an organized method for comprehending these metabolites. Researchers frequently find that classifying ANPs according to their chemical makeup—which is frequently the same as their plant family—helps in their studies. Each species of plant produces a variety of secondary metabolites, but they often specialize in creating specific protective compounds. An example of this is that plants belonging to the Fabaceae family are known to secrete (iso)flavonoids, but plants in the Solanaceae and Malvaceae families are more likely to produce terpenoid phytoalexins. Glucosinolates are a special class of defense chemicals found only in plants in the Brassicales order. It is clear how different plants have their special chemical arsenal to protect themselves [11]. We will discuss secondary metabolites in this chapter, which are thought to aid plants in nematode defense. These are substances that are present in tissues that have been parasitized by nematodes. They may be continuously present in the plant or may be created in reaction to an infection. A correlation appears to exist between these compounds and resistance to nematodes. Plant extracts include a wide variety of nematicidal chemicals, however, not all of them are involved in interactions between plants and nematodes [12].

Phenolic Compounds

Several PPNs like *Meloidogyne incognita* are usually controlled by a large proportion of many groups of phytochemicals classified as phenolic compounds mostly oriented from the phenylpropanoid pathway (Table **1**). They are regarded as fundamental components in conferring resistance against diseases and pests in plants [13]. Following decades of rigorous investigation, it has been determined that these naturally occurring phytochemicals have a significant role in nematode resistance. Numerous recent research studies have emphasized that higher basal along with induced levels of phenolic compounds are associated with improved resistance to nematodes across various plant-nematode interactions [14 - 16]. L-phenylalanine and, to a lesser extent, L-tyrosine are aromatic amino acids that constitute the primary source of phenolic compounds in plants. The enzyme phenylalanine ammonia-lyase (PAL) deamines L-phenylalanine to produce (E)-cinnamic acid. Cinnamic acid-4-hydroxylase (C_4H) can then hydroxylate this to produce para-coumaric acid. Alternatively, tyrosine ammonia lyase (TAL) can directly create para-coumaric acid by deaminating L-tyrosine. The reactive intermediate para-coumaroyl-CoA is created when acetyl coenzyme A and para-coumaric acid are combined by 4-coumarate-CoA ligase (4CL). The phenylpropanoid pathway (PPP) produces several secondary metabolites, including hydroxycinnamic acids, flavonoids, tannins, diarylheptanoids, and stilbenoids. Though the majority of phenolic compounds are generated by the PPP, some lesser-known phenolic metabolites, such as alkylresorcinols, are synthesized by polyketide metabolism [17].

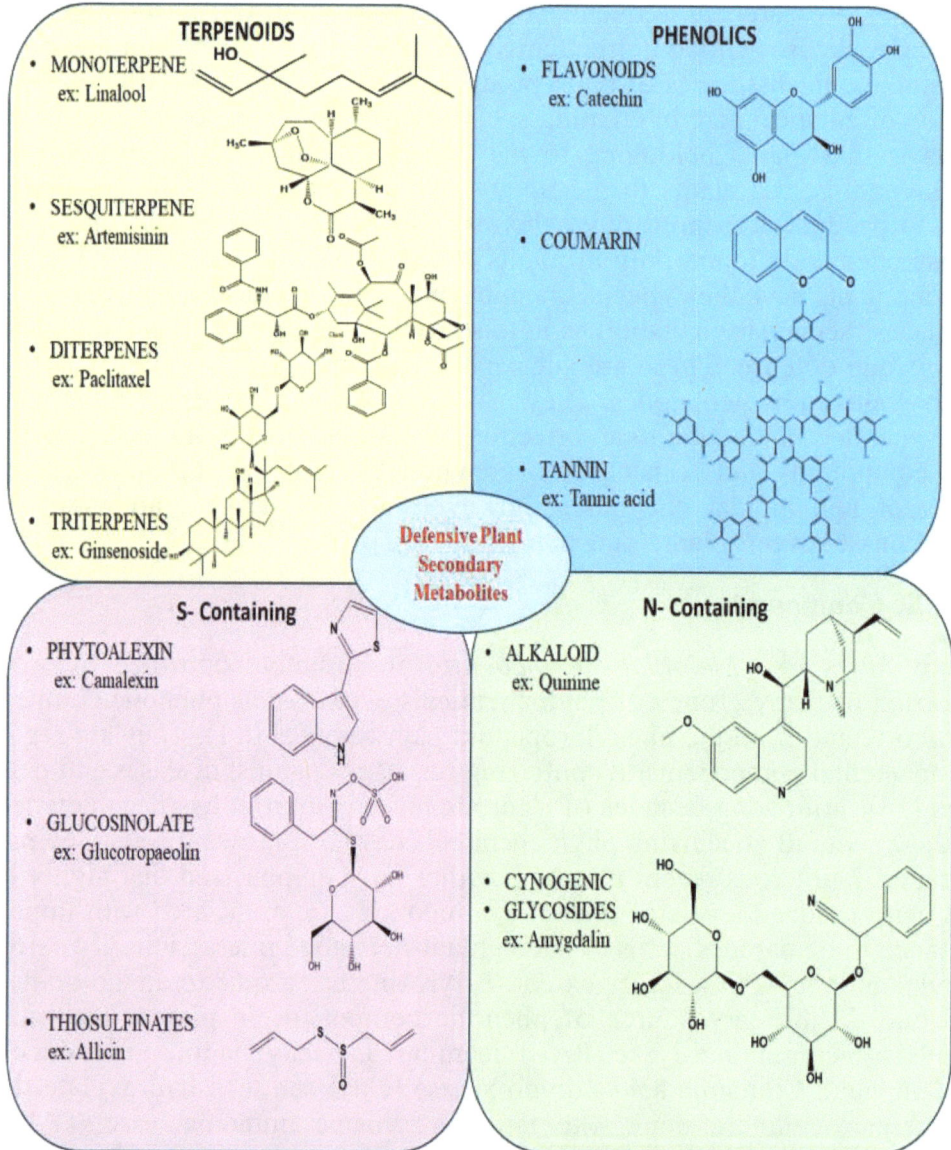

Fig. (1). Chemical structures of some important plant secondary metabolites. (Source: Al-Khayri, *et al.*, 2023: https://doi.org/10.3390/metabo13060716).

Table 1. Plant metabolites with nematicidal activities against *Meloidogyne* species.

S. No.	Plant Species	Part Used	Nematode Type	Nematicidal Constituents	References
1.	*Alstonia boonei*	Leaves	*Meloidogyne incognita*	Fatty acids and triterpenes	[50]
2.	*Armoracia rusticana*	Roots	*Meloidogyne incognita*	Allyl isothiocyanate	[51]
3.	*Azadirachta indica*	Leaves	*Meloidogyne incognita*	Synergy of compounds	[52]
4.	*Azadirachta indica*	Seeds	*Meloidogyne javanica*	Synergy of compounds	[52]
5.	*Berberis brevissima*	Roots	*Meloidogyne javanica*	Berberine	[53]
6.	*Bridelia ferruginea*	Leaves	*Meloidogyne incognita*	Fatty acids and triterpenes	[54]
7.	*Capsicum annuum*	Fruits	*Meloidogyne incognita*	Synergy of compounds	[55]
8.	*Chenopodium ambrosioides*	Leaves	*Meloidogyne incognita*	Essential oil	[56]
9.	*Cymbopogon citratus*	Leaves	*Meloidogyne incognita*	Synergy of compounds	[57]
10.	*Helianthus annuus*	Leaves	*Meloidogyne incognita*	Synergy of compounds	[58]
11.	*Lawsonia inermis*	Leaves	*Meloidogyne incognita*	Synergy of compounds	[59]
12.	*Meliaazedarach*	Fruits	*Meloidogyne incognita*	Synergy of compounds	[60]
13.	*Mentha pulegium*	Aerial parts	*Meloidogyne incognita*	Salicylic acid, pulegone	[61]
14.	*Mentha spicata*	Aerial parts	*Meloidogyne incognita*	Salicylic acid, carvone	[62]
15.	*Myrothecium verrucaria* (fungi)	Whole	*Meloidogyne incognita*	Verrucarin A	[63]
16.	*Parkia biglobosa*	Seeds	*Meloidogyne incognita*	Synergy of compounds	[64]
17.	*Pelargonium graveolens*	Leaves	*Meloidogyne incognita*	Essential oil	[65]

Hydroxycinnamic Acids

The hydroxy forms of (E)-cinnamic acid are known as hydroxycinnamic acids (HAs), and a crucial link in the phenylpropanoid pathway (PPP) is para-coumaric acid. Numerous plants include other HAs, such as caffeic acid, ferulic acid, and sinapic acid, which can be found either as pure compounds or in conjugated forms [18]. Chlorogenic acid was one of the first phenolic compounds to be found for its anti-nematode activity. This molecule, which is an ester of caffeic acid and (-) - quinic acid, accumulates in numerous dicot and monocot plants where plant-parasitic nematodes (PPNs) are infected [13]. Raised levels of chlorogenic acid were associated to nematode resistance in several plant species [19]. *Meloidogyne incognita* galls in a sensitive poplar clone (*Populus tremula* x *Populus alba*) show a dramatically reduced buildup of chlorogenic acid, according to recent investigations. This inhibition of chlorogenic acid production implies a possible pathogenic mechanism used by *M. incognita* [13]. Nevertheless, chlorogenic acid is slightly nematicidal [20]. The idea that chlorogenic acid is a precursor to the uncommon and unstable substance known as nematicidal is one explanation for the disparity, albeit it has not been verified. Among these is quinone, or caffeic acid, which is poisonous to worms [13]. Quinic acid and caffeic acid can be produced by hydrolyzing chlorogenic acid, which can subsequently be oxidized to produce quinone [21]. This hypothesis appears to be supported by the observation that caffeine accumulates after *M. incognita* infection in a resistant tomato cultivar and not in three susceptible ones [22]. Caffeic acid plays a role in lignification, which is another defense reaction against nematodes, thus this remains circumstantial evidence [23]. Moreover, the role of chlorogenic acid in nematode resistance is not uniform; in the relationship between coffee and *Meloidogyne exigua*, both sensitive and resistant cultivars accumulated a comparable amount of chlorogenic acid [24]. It may include salicylic acid (SA), a phenolic plant hormone. While SA has some nematistatic and nematicidal properties in the laboratory, plant roots contain far less of it than is thought to be nematistatic [25]. Rather, SA appears to play a role in nematode resistance by acting like a hormone in plants. SA signaling affects other immune responses against worms and mediates both induced and genetic resistance to root-knot nematodes [26]. Plants that are pre-inoculated with sulfur amino acid (SA) or its chemical analogues are more resistant to nematode infections later on, while mutants lacking SA show greater susceptibility [27].

Stilbenoids and Diarylheptanoids

Stilbenoids and diarylheptanoids are two rather small classes of secondary metabolites that are present in plants and are derived from the PPP. A phenylpropanoyl-CoA and three malonyl-CoA units are coupled during the formation of stilbenoids through a series of condensation events that are facilitated by stilbene synthase. The fundamental C6-C2-C6 stilbene skeleton is formed by this method. Numerous stilbenoid derivatives can be created by further altering these fundamental structures [13]. Though the first steps of diarylheptanoids' biosynthesis are similar to those of stilbenoid biosynthesis, the mechanism itself is not as well understood. The evidence that is now available indicates that stilbenoids and diarylheptanoids have a crucial defensive role in plants that produce them, despite the fact that little research has been done to examine their significance in plant disease resistance [28]. Their role in resistance to plant-parasitic nematodes (PPNs) is also remarkable. Stilbenoids were recently linked to nematode resistance in pine trees and grapevines [29]. One well-known example is the stilbenoid 3-O-methyldihydropinosylvin, which is discovered in the bark and wood of *Pinus strobus* following an infection by the pinewood worm *Bursaphelenchus xylophilus*. This material has potent nematicidal action *in vitro* [30]. The temporal buildup of 3-O-methyldihydropinosylvin in *Pinus strobus* is associated with resistance to *Bursaphelenchus xylophilus*. The nematode can enter the plant successfully at first, but after about a week, it stops moving and reproducing [31]. This is the point in time when the quantities of 3-O-methyldihydropinosylvin in the wood and bark surpass those needed to produce nematicidal activity *in vitro* [13]. *In vitro* tests revealed that all worms were killed after being exposed to 250 µg/ml of 3-O-methyldihydropinosylvin for a full day. The *P. strobus* plants that were infected accumulated about 1,000 and 400 µg/g of the chemical in their bark and wood, respectively. According to these findings, 3-O-methyldihydropinosylvin functions as a phytoalexin and is crucial for *P. strobus'* ability to withstand *B. xylophilus* [32].

While few research have studied the function of stilbenoids and diarylheptanoids in plant disease resistance, existing information suggests that these chemicals act as important defense mechanisms in plants that develop them [33, 34]. Their significance in resistance to plant-parasitic nematodes (PPNs) is also noteworthy. Stilbenoids were involved in nematode tolerance in pine trees and grapevines [29]. The phytoalexins phenolphenalenone are essential for banana resistance against the burying worm *Radopholus similis*. Resistant banana varieties showed significantly higher levels of these cyclic diarylheptanoids near *R. similis* infection sites compared to the susceptible reference cultivar [11]. Among the 13 identified phenylphenalenones, three exhibited significant nematistatic activity *in vitro*. Anigorufone, the most abundant phenylphenalenone, had an IC50 of 59 µg/ml and 23 µg/ml on R. similis motility after 24 and 72 hours, respectively. Anigorufone creates lipid complexes in the worm, resulting in the production of huge lipid-anigorufone droplets and eventually nematode death [11, 13]. In planta, banana roots accumulated approximately 39 mg/g of anigorufone around *R. similis* infection sites, indicating biologically relevant concentrations [29].

Flavonoids

With over 10,000 identified members, flavonoids are a large family of phenolic secondary metabolites that have long been linked to plant resistance to pests and diseases [35]. They are also widely studied in relation to PPN resistance (Fig. **2**). Flavonoid biosynthesis begins resembling to the stilbenoids, with chalcone synthase catalysing the first committed step instead of stilbene synthase [36]. This enzyme combines para-coumaroyl-CoA with three malonyl-CoA units to generate a chalcone skeleton, which is subsequently isomerized to produce the desired flavonoid. Further processing of flavonoids includes hydroxylation, methylation, prenylation, and glycosylation [37]. Flavonoids are categorized into bioflavonoids, isoflavonoids, and neoflavonoids based on their backbone structures [38]. *In vitro* studies have shown that numerous common flavonoids, such as kaempferol, quercetin, and myricetin, have minimal anti-nematode efficacy. Kaempferol has been discovered to suppress the hatching of *Radopholus similis* eggs, while kaempferol, quercetin, and myricetin have showed repelling characteristics and moderately nematistatic effects (though not nematicidal) on *Meloidogyne incognita* juveniles [13]. *In vitro* studies have shown that numerous common flavonoids, such as kaempferol, quercetin, and myricetin, have minimal anti-nematode efficacy. Kaempferol has been discovered to suppress the hatching of *Radopholus similis* eggs, while kaempferol, quercetin, and myricetin have showed repelling characteristics and moderately nematistatic effects (though not nematicidal) on *Meloidogyne incognita* juveniles [39].

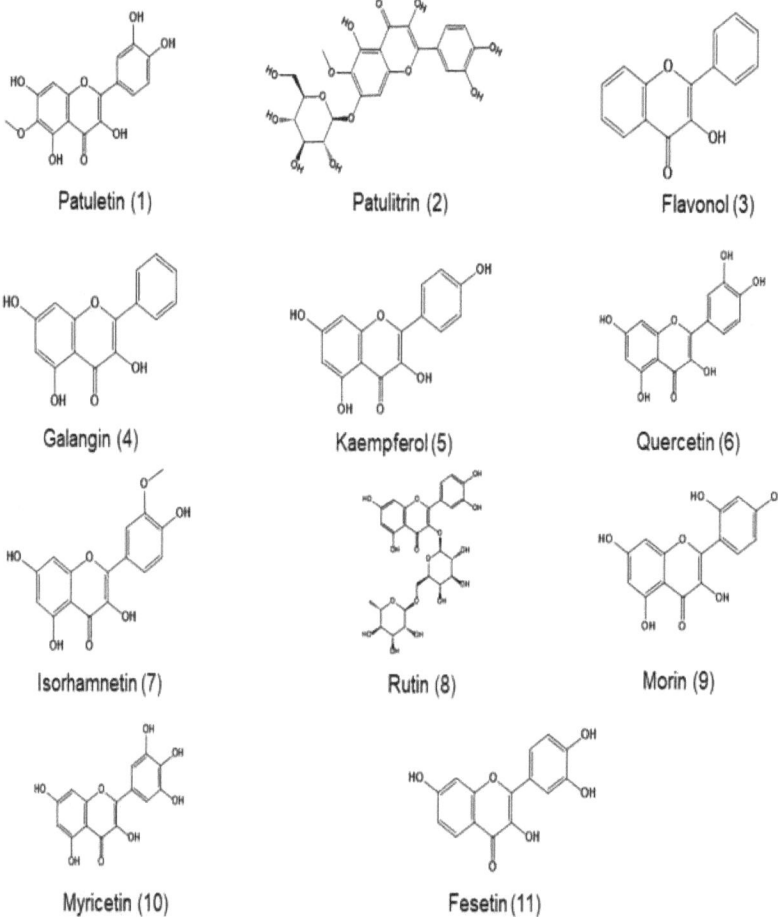

Fig. (2). Depicts the structure of some important flavonoids which have shown promising nematicidal activity (Source: Bano *et al.*, 2020). https://doi.org/10.1007/s10658-020-01988-w.

The Fabaceae family comprises the most widely researched flavonoids in plant-nematode interactions, including isoflavonoids and pterocarpans that are formed in response to infection. Glyceollin I is a pterocarpan that has been extensively investigated [13], *Glycine max* is a phytoalexin generated by soybeans. Glyceollin I accumulates in the head portion of soybean cyst nematodes (*Heterodera glycines*) as early as 8 hours after penetration in a resistant soybean cultivar, and not in a susceptible one [40]. It is unnoticeable in roots before nematode infection, but steadily accumulates subsequently, peaking 4-6 days after penetration and then falling. In the resistant cultivar, glyceollin I levels reached 23 µg/g of fresh root, while the susceptible cultivar accumulated three times less and exhibited no preference for accumulation near the nematode head [41]. The preferential deposition near the nematode head suggests an elicited response, although the molecular patterns in the head region responsible for this remain unidentified [42]. Numerous studies have examined the impact of flavonoids on the interactions between nematodes and grains. Five days after being infected with the stem nematode *Ditylenchus angustus,* shoots of several types of rice (*Oryza sativa)* were collected. Rather than any of the susceptible varieties, a resistant rice variety exhibited 13 µg/g of fresh weight of the flavonoid phytoalexin sakuranetin [43]. However, it is crucial to highlight that this conclusion is only correlated, as the consequences of sakuranetin on *D. angustus* were not studied.

When *Heterodera avenae* or *Pratylenchus neglectus* infect oat (*Avena sativa*), the quantity of three methanol-soluble chemicals by UV-absorbance spectra similar to flavonoids increases two to three times in the shoot and root [44]. Similarly, its concentration is similarly increased by foliar administration of the defense hormone methyl jasmonate. A crude methanol extract of methyl jasmonate-induced oat showed a high nematicidal effect with respect to *H. avenae.* Two of the three inducible flavonoid phytoalexins were successfully purified, and they all showed notable nematicidal properties. These inducible flavonoids were found to be luteolin-C-hexoside-O-pentoside (could not be isolated), O-methyl-apigenin-C-deoxyhexoside-O-hexoside (nematicidal), and apigenin-C-hexoside-O-pentoside (not nematicidal). These flavone-C- glycosides were characterized as [45]. Ten days after inoculation, pre-treatment with methyl jasmonate dramatically decreased the overall nematode population for both *P. neglectus* and *H. avenae,* and increased the proportion of nematodes outside the root as opposed to inside. When susceptible wheat plants were given an extract rich in flavonoids derived from induced oat plants, a comparable result was seen [46]. These findings suggest that oat-derived inducible flavonoids have nematicidal and repellant properties. But only when these flavonoids are present before or soon after penetration (for example, by methyl jasmonate pre-treatment) do they work effectively in planta against *H. avenae* and *P. neglectus*. The quantity of inducible flavonoids in untreated susceptible plants ultimately increased to the level observed in methyl jasmonate-induced plants upon infection by *H. avenae* or *P. neglectus*, but the nematodes were still able to proliferate [47]. Flavonoid production is disrupted in *Arabidopsis thaliana* transparent testa (tt) mutants, that are used to investigate the role of flavonoids in PPN resistance. According to one investigation, the susceptibility of all examined tt mutants to *M. incognita* was the same as that of their wild type [48]. However, in contrast to their wild kinds, the majority of tt mutants demonstrated either unchanged or slightly enhanced susceptibility to *Heterodera schachtii*, according to another study. These results imply that flavonoids contribute very little to *A. thaliana's* PPN resistance [49].

Several researchers believe that sedentary nematodes may employ flavonoids as part of their detrimental activity. This hypothesis is based on evidence that plant-parasitic nematodes (PPN) impair plant auxin homeostasis during feeding site development, and that several flavonoids have been identified as auxin transport inhibitors. However, the evidence for this theory remains circumstantial. If PPN dramatically affected flavonoids as a pathogenic strategy, it would be expected that *Arabidopsis thaliana* mutants would show higher vulnerability, which is not always the case [66]. Regarding the notion that PPN take advantage of flavonoids, it has been documented that in developing *Meloidogyne incognita* feeding sites in white clover, the expression of chalcone synthase, a critical gene in flavonoid biosynthesis, correlates spatiotemporally with an increased auxin response. Furthermore, fewer *Meloidogyne incognita* galls than the wild type are seen in a transgenic strain of *Medicago truncatula* that lacks flavonoids. However, these findings are just suggestive and do not establish beyond a reasonable doubt that PPN modifies auxin through flavonoids.

The information provided in this section together indicates that the participation of flavonoids in plant-parasitic nematode (PPN) resistance may depend on the particular flavonoids and nematodes involved, as well as the possible timing of flavonoid accumulation during the infection process. Although there is compelling evidence linking the flavonoid glyceollin I to soybean resistance to Heterodera glycines and *Meloidogyne incognita*, there is conflicting data from different pathosystems [13].

Tannins

Terpenoids, which comprise terpenes and their forms, are one of the most broad groups of secondary metabolites of plants, with over 60,000 compounds identified. Terpenes are produced through the condensation of two or more activated isoprene units (C5 building blocks): isopentenyl pyrophosphate or its isomer dimethylallyl pyrophosphate [67]. Depending on the amount of C5 units are involved, this condensation can produce a C10 (monoterpene), C15 (sesquiterpene), or C20 (diterpene) terpene. Sesqui- and diterpene units can then undergo head-to-head condensation to generate C30 (triterpenes, such as sterols) or C40 units (tetraterpenes, such as carotenoids) [68]. Terpenoids are formed by further substituting all terpenes, such as through hydroxylation or acetylation. Terpenoids are an extremely diverse family of secondary metabolites because of the numerous and frequently promiscuous enzymes that act on them [69].

In earlier times, it was thought that tannins prevented herbivores from grazing by causing protein precipitation. Recent research, however, indicates that this impact might not be significant *in vivo* and that tannins instead get their active ingredients from cytotoxic and antinutritive chemicals that are created when plant polyphenol oxidases or the alkaline stomach environments of many insect herbivores oxidize tannins [70].

Additionally, it has been proposed that tannins contribute to resistance against the pinewood worm *B. xylophilus*. When *B. xylophilus* was grown on the phloem sap of eight distinct pine species, the amount of condensed tannins was found to be negatively correlated with the bacteria's growth rate [71]. However, it was also shown that there was a negative correlation between the rate of nematode growth and the concentrations of total phenolic compounds and total flavonoids. This makes it difficult to assess the relative contributions of condensed tannins, flavonoids, and other phenolic metabolites to the inhibitory effect on *B. xylophilus* [13].

Terpenoids

Terpenoids, which include terpenes and their forms, are among the most broad category of plant secondary metabolites, with over 60,000 chemicals discovered. Terpenes are generated by the condensation of two or more activated isoprene units (C5 building blocks): isopentenyl pyrophosphate or its isomer dimethylallyl pyrophosphate [72]. Depending on the amount of C5 units are involved, this condensation can produce a C10 (monoterpene), C15 (sesquiterpene), or C20 (diterpene) terpene. Sesqui- and diterpene units can then undergo head-to-head condensation to generate C30 (triterpenes, such as sterols) or C40 units (tetraterpenes, such as carotenoids) [73]. Terpenoids are formed by further substituting all terpenes, such as through hydroxylation or acetylation. Terpenoids are an extremely diverse family of secondary metabolites because of the numerous and frequently promiscuous enzymes that act on them [74].

A developing arms race with these adversaries may have played a significant role in the increasing terpenoid variety observed throughout plant evolution, as many terpenoids are active against pests and diseases [75]. In plant-nematode interactions, the terpenoid aldehydes (TAs) present in cotton (*Gossypium* sp.)—which include gossypol and its derivatives—are the most thoroughly researched category of terpenoids. Despite being a polyphenolic chemical, gossypol is classified as a terpenoids because of the way it is made: it is created by oxidatively combining two sesquiterpene units that have been continuously oxidized [76]. Pepper (*Capsicum annuum*) was shown to be susceptible to *M. incognita* when the relative amounts of various terpenes in root exudates of various kinds were found to be connected with this vulnerability. Multiple terpenes generated by *C. annuum* were shown to either repel or attract *M. incognita* J2s, according to olfactometer investigations. This implies that released terpenes may either aid or hinder host-finding, therefore increasing vulnerability to PPN [77].

It is unknown whether the various sesquiterpene phytoalexins from the Solanaceae family contribute to plant-parasitic nematode (PPN) resistance. According to one study, potato cultivars with stronger resistance to *Globodera rostochiensis* are those in which the sesquiterpene solavetivone makes up a larger than usual fraction of total sesquiterpene levels [13]. Nevertheless cultivars with high production of solavetivones all had a similar ancestor line, *Solanum tuberosum* ssp. *andigena*, indicating that resistance in these lines might not be a result of accumulation of solavetivones but rather of another feature acquired from this ancestor line [78]. Terpenoid phytoalexins, which are produced by a number of plant species, are not yet known to play a part in nematode resistance. For instance, maize (*Zea mays*) yields phytoalexins that are sesquiterpenoid (zealexins) and diterpenoid (kauralexins and dolabralexins) [79], Rice produces three distinct types of diterpenoid phytoalexins: momilactones, phytocassanes, and oryzalexins. These terpenoids help defend against fungal, bacterial, and insect pests and infections [80]. Treatment with resistance inducers that lessen sensitivity to nematodes can induce them [81].

ALKALOIDS IN PLANT RESISTANCE TO NEMATODES

Alkaloids are a class of secondary metabolites found in plants that are distinguished by heterocyclic rings that include nitrogen. Alkaloids are notable for their broad distribution across plant species and their strong defense qualities against pests and infections, despite their complicated taxonomy [82].

Alkaloids

Tomato (*Solanum lycopersicum)* contains α-tomatine, a well-known alkaloid that helps with insect and fungus resistance. Studies have revealed no link between basal root α-tomatine concentrations and resistance levels against the root-knot nematode *M. incognita* in twelve tomato cultivars [83]. Additionally, during nematode infection, the quantities of α-tomatine did not alter in cultivars that were resistant or susceptible, indicating that it is unlikely to play a significant role in tomato defense against *M. incognita*. Similarly, no relationship was discovered between the glycoalkaloid content of cultivated potatoes (*Solanum tuberosum*), which includes α-solanine and α-chaconine, and resistance to the potato cyst nematodes *G. pallida* and *G. rostochiensis* [84]. Even in lines created by crossing cultivated potato with the *G. pallida*-resistant wild potato *Solanum vernei*, there was no consistent link between glycoalkaloid concentration and nematode resistance. These findings imply that glycoalkaloids are not the main contributors to resistance against potato cyst nematodes [85]. Camalexin is a well-known indole alkaloid phytoalexin found in the model plant *Arabidopsis thaliana* that is linked to resistance to a variety of pests and illnesses. Mutants that produce less camalexin—such cyp79b2/b3 and pad3—are more vulnerable to nematodes like *Meloidogyne incognita* and *Heterodera schachtii* [86]. This underscores the importance of camalexin in *Arabidopsis* defense responses against nematodes.

Saponins

Saponins, another type of plant secondary metabolite with surfactant qualities, have also been studied for their possible role in nematode resistance. There was no link between resistance to *Meloidogyne hapla* and *Ditylenchus dipsaci* and basal saponin concentrations in six varieties of alfalfa (*Medicago sativa*). This implies that saponins might not be very important in mediating alfalfa's resistance to these worms [13]. In contrast, oats (*Avena sativa*) have indicated possible linkages between saponins and resistance to the cereal cyst worm *Heterodera avenae*. The analysis of root tip extracts from oat lines with various resistance levels revealed multiple saponin peaks that were substantially linked with resistance. These saponins were identified as avenacins, which are renowned for their antibacterial and perhaps anti-nematode effects in oats [13].

Benzoxazinoids

Studies on benzoxazinoids in wheat (*Triticum aestivum*) have revealed contrasting effects on resistance to nematodes. Research by Hama *et al.*, in wheat infected with an arbuscular mycorrhizal fungus that inhibited benzoxazinoid synthesis was shown to be more vulnerable to *Pratylenchus neglectus*. Furthermore, wheat cultivars that were more susceptible to *P. neglectus* had lower basal and induced amounts of benzoxazinoids. Specifically, HMBOA-glucoside and HDMBOA-glucoside correlated positively with resistance to *P. neglectus*, whereas DIBOA-glucoside did not. These results suggest that certain benzoxazinoids have unique roles in wheat resistance to nematodes [87].

Glucosinolates

Glucosinolates, which are typical secondary metabolites of plants in the Brassicales order, are known for their function in pest and disease defense by producing biocidal isothiocyanates when hydrolyzed by myrosinase. *In vitro* studies have showed high nematicidal activity of glucosinolates, indicating that biofumigation with *Brassica* seed meal or green manures could be a promising alternative to chemical fumigation [88]. However, their role in nematode resistance *in vivo* remains less explored. In *Brassica napus*, susceptibility to *P. neglectus* was found to be uncorrelated with total glucosinolate content across different accessions [89]. A particular glucosinolate, 2-phenylethyl glucosinolate, demonstrated a clear association with resistance. Cultivars generating over a critical threshold of 8-12 μmol/g fresh root tissue were less sensitive to *P. neglectus*. Interestingly, this trend was not found with total glucosinolate concentration, suggesting that the efficiency of *B. napus* as a biofumigant against *P. neglectus* is directly tied to its 2-phenylethyl glucosinolate content [90].

ORGANOSULFUR COMPOUNDS

Marigolds (*Tagetes* sp.) are known for their resistance to plant-parasitic nematodes, attributed in part to polythienyl compounds such as α-terthienyl present in their roots and exudates [89]. A meta-analysis encompassing 175 species of Asteraceae highlighted a considerable correlation between the polythylene content of these plants and their ability to decrease *Pratylenchus penetrans* populations. Fifteen out of sixteen Asteraceae species that are known to produce α-terthienyl were found to be suppressive against *P. penetrans* [91].

Furthermore, anti-PPN action was found to be associated with an unidentified acetylenic dithio molecule that had a characteristic red color. Out of the 12 assessed Asteraceae plants that release this chemical, 11 of them demonstrated inhibition of *P. penetrans*. These findings underscore the significant role of organosulfur compounds, specifically polythylenes and acetylenic dithio compounds, in mediating the resistance of Asteraceae plants against plant-parasitic nematodes. The production of α-terthienyl and related chemicals is not fully known, despite several hypothesized methods [92]. In contrast, the nematicidal mode of action of α-terthienyl has been extensively investigated compared to most other ANPs. It has been demonstrated that α-terthienyl produces reactive oxygen species when activated by light or peroxidase. Studies on Caenorhabditis elegans have shown that RNAi lines with altered levels of superoxide dismutase and glutathione peroxidase exhibit increased or decreased susceptibility to α-terthienyl, respectively. Additionally, α-terthienyl effectively penetrates the nematode hypodermis, indicating its nematicidal effect through the induction of oxidative stress within the nematode.

Asparagus (*Asparagus officinalis*), renowned for its nematode-suppressive properties, produces a highly nematicidal compound in its roots identified as asparagusic acid [93]. Asparagusic acid demonstrates potent nematicidal properties against several plant-parasitic nematode (PPN) species and inhibits Heterodera egg hatching at 50 ppm [94]. With asparagus roots containing at least 35 ppm of asparagusic acid, this compound likely plays a prominent role as a phytoanticipin in the anti-nematode activity of asparagus plants [89]. The biosynthesis of asparagusic acid, unique to asparagus, involves isobutyric acid and methacrylic acid as precursors, with cysteine serving as the sulfur donor.

Challenges in the Identification of ANPs

Despite technical advancements, ANP identification is still difficult. Traditionally, the process of identifying ANPs involved making crude extracts from plants that suppress nematodes, fractionating them, and assessing the anti-nematode activity of the active fractions. Analytical techniques such as elemental analysis, UV/VIS and infrared spectroscopy, color reagents, and, more recently, mass spectrometry (MS) or nuclear magnetic resonance (NMR) spectroscopy made it easier to propose candidate compounds that were subsequently manufactured and used as standards [93]. In several cases, metabolites already implicated in plant resistance to other pests or pathogens were studied for their function in PPN resistance, which sped up the identification process. This targeted method enabled the finding of glyceollin I as critical in soybean PPN resistance [95]. Histopathological methods provided further evidence by showing preferential accumulation of metabolites near PPN infection sites, such as glyceollin I in resistant soybeans against *H. glycines* [96].

A recent method is examining mutants that are deficient in metabolite biosynthesis in order to establish causal linkages to resistance. However, creating mutants is time-consuming and necessitates detailed knowledge of biosynthetic routes or random mutagenesis, followed by screening. Mutant analysis has successfully revealed the involvement of secondary metabolites in nematode resistance in model plants such as *Arabidopsis thaliana* [97].

Untargeted metabolomics is a viable option since it provides a full picture of metabolite profiles in plants with variable susceptibility to PPNs. This technique involves low prior knowledge and negligible amounts of input material, although later extraction and purification of ANPs for bioassays may still require higher quantities. Recent research employing untargeted metabolomics and other approaches has revealed possible biomarkers and candidate ANPs in plant-nematode interactions. For example, gas chromatography-mass spectrometry (GC-MS) found significant metabolic alterations in bermudagrass lines with different nematode susceptibility, indicating possible ANPs such as L-pipecolic acid and phenylalanine. Similarly, combined GC-MS and transcriptomics on soybeans exposed to H. glycines identified metabolites like 4-vinylphenol and piperine with *in vitro* nematicidal [98]. Overall, advancements in metabolomics have enhanced our ability to identify and understand the roles of ANPs in plant-nematode interactions, offering insights into novel strategies for nematode control in agriculture.

NMR has been employed as a stand-alone method in numerous studies on plant-nematode interactions. Researchers, for example, used NMR and UV/VIS spectrophotometric tests to profile the roots of *Meloidogyne exigua*-susceptible and resistant coffee cultivars at different time points after worm infection. They discovered a higher accumulation of phenolic chemicals, sucrose, and fumaric acid in the resistant cultivar, with quinic acid being significantly more plentiful than in the susceptible one. The levels of amino acids, total carbs, and alkaloids have no association with resistance [99]. In another investigation, untargeted NMR-based metabolomics was used to analyze the tomato-*Meloidogyne incognita* relationship. Root tissue from four tomato cultivars (two extremely sensitive and two highly resistant) were studied 38 days post-inoculation, finding considerably greater levels of metabolites including caffeic acid and glucose in the resistant accessions post-infection, indicating a potential involvement in resistance. NMR metabolomics was also employed in studying nematode interactions in soursop tree (*Annona muricata*) seedlings, known for their high nematode resistance. Root extracts from soursop seedlings inoculated with *M. javanica* were analyzed over different time points using NMR, followed by bio-assay-guided fractionation. The chloroform fraction showed nematistatic activity, with NMR analysis identifying acetogenins as the active compounds. The study indicated minimal metabolomic shifts in nematode-inoculated plants beyond 2 days post-inoculation, highlighting their high resistance to nematode invasion.

Despite the effectiveness of untargeted metabolomics in discovering potential ANPs, metabolite identification remains difficult due to the large number of secondary metabolites in plants and the biases induced by various analytical methodologies. While NMR provides detailed structural information and is highly repeatable, its sensitivity and spectral overlap are limited, making metabolite identification difficult. Strategies such as 2D NMR, LC-NMR coupling, or fractionation prior to analysis have been investigated to address these problems in metabolomics studies [100].

GC-MS and HPLC-MS are popular metabolomics techniques, each with unique strengths and limitations. GC-MS employs high-energy ionization to produce comprehensive and reproducible MS spectra that can be compared to biological databases. However, it can only identify volatile or derivatized metabolites, which limits its usefulness for smaller molecules. HPLC-MS, on the other hand, may study non-volatile chemicals but is limited by mass spectra repeatability and matrix effects such as adduct formation. Despite these obstacles, HPLC-MS is adept in analyzing semi-polar metabolites using a common reversed-phase C18 column, which is prevalent in studies of plant-nematode interactions. Choosing the appropriate metabolomics technique involves strategic planning, often guided by transcriptome analysis to identify RNA-seq revealed significant modulation of phenolic metabolism genes upon *M. incognita* infection, prompting the selection of HPLC-MS/MS to analyze these metabolites [101].

Despite methodological limitations, metabolomics techniques provide promise for understanding the roles of secondary metabolites in nematode resistance. The studies discussed in this review demonstrate how untargeted metabolomics might uncover possible novel antinematodal phytochemicals (ANPs), even in non-model plants like *Annona muricata* and *Cynodon transvaalensis*. Applications of ANPs extend beyond scientific discovery to practical uses such as developing new botanical nematicides. For decades, compounds like isothiocyanates have been utilized in chemical and biofumigation approaches for nematode control [102]. Novel ANPs discovered using metabolomics could be promising leads for building next-generation phytochemical-based nematicides. Furthermore, ANPs discovered in metabolomics investigations represent prospective targets for crop improvement *via* genetic engineering. Although still theoretical, genetic manipulation could improve crop resistance to nematodes by increasing native ANP production or adding novel ANP biosynthetic pathways.

Additionally, ANPs are used as indicators in worm resistance breeding efforts. High-throughput targeted metabolomics has the potential to replace labor-intensive nematode resistance experiments by screening breeding lines for specific metabolic indicators linked to resistance mechanisms. Despite the variability of resistance mechanisms within breeding populations, the use of metabolic markers shows promise for accelerating the creation of nematode-resistant crops [103].

Metabolic markers also have the potential for screening-induced resistance, where exogenously applied chemicals or microbes stimulate plant immune responses against pests or pathogens, including PPNs [104]. By reliably correlating metabolic markers with induced resistance in specific plant-PPN systems, researchers can screen for novel resistance inducers and study the longevity of induced resistance more efficiently than through traditional inoculation experiments [105]. Improvements in untargeted metabolomics, when paired with high-dimensional statistical approaches, offer the potential to allow selection based on metabolic profiles rather than specific markers. This method evaluates many features, each of which could indicate undiscovered compounds related to resistance to PPNs, allowing resistance to be predicted simultaneously. While effectively used in breeding for resistance to fungal diseases, this strategy remains largely unexplored in the context of plant-nematode interactions. However, cases like oat *H. avenae* interactions mentioned earlier suggest its potential utility, highlighting the need for comprehensive metabolic profiling across various metabolite classes, including flavonoids and saponins.

CONCLUSION AND FUTURE PERSPECTIVES

Since the mid-20th Century, extensive research has underscored plants' reliance on small molecules to combat PPN. Yet, the vast diversity among metabolites, plants, and nematode species complicates generalizations from existing literature. Firstly, few ANPs have definitively proven their causal role in nematode resistance. Establishing causality typically demands demonstrating; (1) Higher compound abundance in resistant varieties, (2) *In vitro* anti-nematode activity of purified/synthesized ANPs, (3) *In planta* accumulation at relevant sites, and (4) Diminished resistance upon ANP production inhibition. While some studies meet the first three criteria, very few achieve the fourth, often due to pre-omics era limitations [106]. Secondly, many cited studies are dated, averaging nearly 25 years old, with recent reviews less focused on ANPs. Revitalizing ANP research *via* contemporary methods (*e.g.*, targeted mutagenesis, transcriptomics, metabolomics) holds promise for advancing plant nematology [104].

Thirdly, studies reveal that the total concentrations of certain secondary metabolite classes often fail to correlate with PPN resistance. Instead, specific, sometimes low-abundance metabolites within these classes show strong associations, underscoring the need for high-resolution analytical techniques like HPLC- or GC-MS [107]. Lastly, a methodological shift is evident over time: from targeted searches for ANPs in extracts of known nematode-suppressive plants, towards more comprehensive, -omics-era approaches. Despite their current scarcity, recent metabolomics studies have hinted at novel ANPs in various plants, suggesting a potential resurgence in ANP research with expanding biological databases and user-friendly data analysis tools [108].

COMPETING INTEREST

The authors state that they have no known financial conflicts of interest or close connections that could have influenced the findings of the study presented in this work.

ACKNOWLEDGMENTS

The authors express their gratitude for the support and resources provided by Chandigarh University, Mohali, India.

REFERENCES

[1]   Castro-Moretti FR, Gentzel IN, Mackey D, Alonso AP. Metabolomics as an emerging tool for the study of plant–pathogen interactions. Metabolites 2020; 10(2): 52.
[http://dx.doi.org/10.3390/metabo10020052] [PMID: 32013104]

[2]   Roopa K, Gadag AS. Importance of biopesticides in the sustainable management of plant-parasitic nematodes. In: Ansari RA, Rizvi R, Mahmood I, Eds., Management of phytonematodes: Recent advances and future challenges 2020; pp. 205-27.
[http://dx.doi.org/10.1007/978-981-15-4087-5_9]

[3]   Kantor C, Eisenback JD, Kantor M. Biosecurity risks to human food supply associated with plant-parasitic nematodes. Front Plant Sci 2024; 15: 1404335.
[http://dx.doi.org/10.3389/fpls.2024.1404335] [PMID: 38745921]

[4]   Phani V, Khan MR, Dutta TK. Plant-parasitic nematodes as a potential threat to protected agriculture: Current status and management options. Crop Prot 2021; 144: 105573.
[http://dx.doi.org/10.1016/j.cropro.2021.105573]

[5]   Sorribas FJ, Djian-Caporalino C, Mateille T. Nematodes. In: Gullino ML, Albajes R, Nicot PC, Eds., Integrated pest and disease management in greenhouse crops 2020; pp. 147-74.

[6]   Sivasubramaniam N, Hariharan G, Zakeel MCM. Sustainable management of plant-parasitic nematodes: an overview from conventional practices to modern techniques. In: Ansari RA, Rizvi R, Mahmood I, Eds., Management of phytonematodes: Recent advances and future challenges 2020; pp. 353-99.

[7]   Engelbrecht G, Claassens S, Mienie CMS, Fourie H. South Africa: an important soybean producer in sub-Saharan Africa and the quest for managing nematode pests of the crop. Agriculture 2020; 10(6): 242.
[http://dx.doi.org/10.3390/agriculture10060242]

[8]   Rajwade JM, Chikte RG, Paknikar KM. Nanomaterials: new weapons in a crusade against phytopathogens. Appl Microbiol Biotechnol 2020; 104(4): 1437-61.
[http://dx.doi.org/10.1007/s00253-019-10334-y] [PMID: 31900560]

[9]   Menéndez-Pedriza A, Jaumot J. Interaction of environmental pollutants with microplastics: a critical review of sorption factors, bioaccumulation and ecotoxicological effects. Toxics 2020; 8(2): 40.
[http://dx.doi.org/10.3390/toxics8020040] [PMID: 32498316]

[10]   Roy C, Singh RS, Yadav A, Kumar S, Kumari M. Secondary metabolites: evolutionary perspective, *in vitro* production, and technological advances. In; Siddiqui MW, Bansal V, Prasad K Eds., Plant secondary metabolites, Volume Two 2017; pp. 27-52.

[11]   Koche D. Role of secondary metabolites in plants' defense mechanism. MakdeKH Glimpse Curr Vistas Plant Sci Res Hislop Coll Publ Cell Nagpur 2014; pp. 1-16.

[12]   Mwamula AO, Kabir MF, Lee D. A review of the potency of plant extracts and compounds from key families as an alternative to synthetic nematicides: history, efficacy, and current developments. Plant Pathol J 2022; 38(2): 53-77.
[http://dx.doi.org/10.5423/PPJ.RW.12.2021.0179] [PMID: 35385913]

[13]   Desmedt W, Mangelinckx S, Kyndt T, Vanholme B. A phytochemical perspective on plant defense against nematodes. Front Plant Sci 2020; 11: 602079.
[http://dx.doi.org/10.3389/fpls.2020.602079] [PMID: 33281858]

[14]   Hammad E, El-Deriny M, Ibrahim D. Efficiency of humic acid and three commercial biocides against *Meloidogyne incognita* and *Tylenchulus semipenetrans* associated with olive plants. Egypt J Phytopathol 2021; 49(2): 103-15.
[http://dx.doi.org/10.21608/ejp.2021.108638.1049]

[15]   Lopes-Caitar VS, Nomura RBG, Hishinuma-Silva SM, *et al.* Time course RNA-seq reveals soybean responses against root-lesion nematode and resistance players. Plants 2022; 11(21): 2983.
[http://dx.doi.org/10.3390/plants11212983] [PMID: 36365436]

[16]   Shah I, Shah MA, Nawaz MA, Pervez S, Noreen N, Vargas-de la Cruz C, *et al.* Analysis of other phenolics (capsaicin, gingerol, and alkylresorcinols). In: Siddiqui MW, Bansal V, Prasad K, Silva AS, Nabavi SF, Saeedi M, Nabavi SM, Eds, Recent advances in natural products analysis. Elsevier 2020; pp. 255-71.
[http://dx.doi.org/10.1016/B978-0-12-816455-6.00006-8]

[17]   Abd-Elgawad MMM. Understanding molecular plant–nematode interactions to develop alternative approaches for nematode control. Plants 2022; 11(16): 2141.
[http://dx.doi.org/10.3390/plants11162141] [PMID: 36015444]

[18]   Demuth T. Characterization of processing-induced arabinoxylan modifications and their effect on the *in vitro* fermentation by gut microbiota, 2020;

[19]   Oosterbeek M, Lozano-Torres JL, Bakker J, Goverse A. Sedentary plant-parasitic nematodes alter auxin homeostasis *via* multiple strategies. Front Plant Sci 2021; 12: 668548.
[http://dx.doi.org/10.3389/fpls.2021.668548] [PMID: 34122488]

[20]   Mukhamedsadykova AZ, Kasela M, Kozhanova KK, *et al.* Anthelminthic and antimicrobial effects of hedge woundwort (*Stachys sylvatica* L.) growing in Southern Kazakhstan. Front Pharmacol 2024; 15: 1386509.
[http://dx.doi.org/10.3389/fphar.2024.1386509] [PMID: 38769997]

[21]   Zieniuk B. Dihydrocaffeic acid—is it the less known but equally valuable phenolic acid? Biomolecules 2023; 13(5): 859.
[http://dx.doi.org/10.3390/biom13050859] [PMID: 37238728]

[22]   Al-Khayri JM, Rashmi R, Toppo V, *et al.* Plant secondary metabolites: The weapons for biotic stress management. Metabolites 2023; 13(6): 716.
[http://dx.doi.org/10.3390/metabo13060716] [PMID: 37367873]

[23]   Yan Y, Mao Q, Wang Y, *et al. Trichoderma harzianum* induces resistance to root-knot nematodes by increasing secondary metabolite synthesis and defense-related enzyme activity in *Solanum lycopersicum* L. Biol Control 2021; 158: 104609.
[http://dx.doi.org/10.1016/j.biocontrol.2021.104609]

[24]   Santra HK, Banerjee D. Natural products as fungicide and their role in crop protection. In: Singh J, Yadav AN, Eds., Natural bioactive products in sustainable agriculture 2020; pp. 131-219.
[http://dx.doi.org/10.1007/978-981-15-3024-1_9]

[25]   Kisiriko M, Anastasiadi M, Terry LA, Yasri A, Beale MH, Ward JL. Phenolics from medicinal and aromatic plants: Characterisation and potential as biostimulants and bioprotectants. Molecules 2021; 26(21): 6343.
[http://dx.doi.org/10.3390/molecules26216343] [PMID: 34770752]

[26] Ojeda-Rivera JO, Ulloa M, Roberts PA, *et al.* Root-knot nematode resistance in *Gossypium hirsutum* determined by a constitutive defense-response transcriptional program avoiding a fitness penalty. Front Plant Sci 2022; 13: 858313.
[http://dx.doi.org/10.3389/fpls.2022.858313] [PMID: 35498643]

[27] Yokotani N, Hasegawa Y, Sato M, *et al.* Transcriptome analysis of *Clavibacter michiganensis* subsp. michiganensis-infected tomatoes: a role of salicylic acid in the host response. BMC Plant Biol 2021; 21(1): 476.
[http://dx.doi.org/10.1186/s12870-021-03251-8] [PMID: 34666675]

[28] Ali O, Ramsubhag A, Daniram Benn Jr Ramnarine S, Jayaraman J. Transcriptomic changes induced by applications of a commercial extract of *Ascophyllum nodosum* on tomato plants. Sci Rep 2022; 12(1): 8042.
[http://dx.doi.org/10.1038/s41598-022-11263-z] [PMID: 35577794]

[29] Hajam YA. Diksha, Kumar R, Lone R. In: Lone R, Khan S, Al-Sadi AM, Eds., Plant phenolics and their versatile promising role in the management of nematode stress. Plant Phenolics in Biotic Stress Management. Springer 2024; pp. 389-416.
[http://dx.doi.org/10.1007/978-981-99-3334-1_16]

[30] Zhang F, Kajiwara J, Mori Y, Ohira M, Tsutsumi Y, Kondo R. Metabolites from resistant and susceptible *Pinus thunbergii* after inoculation with pine wood nematode, Am J Plant Sci. 2013 Mar;4(3).

[31] Pezhman Z. Xylem formation in *pinus radiata* callus under *In vitro* conditions, 2021.

[32] Menéndez-Gutiérrez M, Matsunaga K, Togashi K. Relationship between pine wilt-tolerance rankings of *Pinus thunbergii* trees and the number of *Bursaphelenchus xylophilus* passing through branch sections. Nematology 2017; 19(9): 1083-93.
[http://dx.doi.org/10.1163/15685411-00003108]

[33] Bian X, Zhao Y, Xiao S, Yang H, Han Y, Zhang L. Metabolome and transcriptome analysis reveals the molecular profiles underlying the ginseng response to rusty root symptoms. BMC Plant Biol 2021; 21(1): 215.
[http://dx.doi.org/10.1186/s12870-021-03001-w] [PMID: 33985437]

[34] Cuéllar-Torres EA, Aguilera-Aguirre S, Hernández-Oñate MÁ, *et al.* Molecular aspects revealed by omics technologies related to the defense system activation in fruits in response to elicitors: A review. Horticulturae 2023; 9(5): 558.
[http://dx.doi.org/10.3390/horticulturae9050558]

[35] Kaur N, Ahmed T. Bioactive secondary metabolites of medicinal and aromatic plants and their disease-fighting properties. In: Aftab T, Hakeem KR, Eds., Medicinal and aromatic plants Healthcare and industrial application 2021; pp. 113-42.
[http://dx.doi.org/10.1007/978-3-030-58975-2_4]

[36] Pandith SA, Dhar N, Rana S, *et al.* Functional promiscuity of two divergent paralogs of type III plant polyketide synthases. Plant Physiol 2016; 171(4): 2599-619.
[http://dx.doi.org/10.1104/pp.16.00003] [PMID: 27268960]

[37] Paiva NL. An introduction to the biosynthesis of chemicals used in plant-microbe communication. J Plant Growth Regul 2000; 19(2): 131-43.
[http://dx.doi.org/10.1007/s003440000016] [PMID: 11038223]

[38] Sangeetha KS, Umamaheswari S, Reddy CUM, Kalkura SN. Flavonoids: Therapeutic potential of natural pharmacological agents. Int J Pharm Sci Res 2016; 7(10): 3924.

[39] Torto B, Kirwa H, Kihika R, Murungi LK. Strategies for the manipulation of root knot nematode behavior with natural products in small scale farming systems. In: Beck JJ, Rering CC, Duke SO, Eds, Roles of natural products for biorational pesticides in agriculture. ACS Publications 2018; pp. 115-26.
[http://dx.doi.org/10.1021/bk-2018-1294.ch009]

[40] Ramatsitsi MN. Mechanism of resistance to *Meloidogyne incognita* and *Meloidogyne javanica* in *Cucumis africanus* and *Cucumis myriocarpus* seedlings, 2017.

[41] Luck AN, Anderson KG, McClung CM, *et al.* Tissue-specific transcriptomics and proteomics of a filarial nematode and its *Wolbachia endosymbiont.* BMC Genomics 2015; 16(1): 920.
[http://dx.doi.org/10.1186/s12864-015-2083-2] [PMID: 26559510]

[42] Zhang L, Zhou Z, Guo Q, *et al.* Insights into adaptations to a near-obligate nematode endoparasitic lifestyle from the finished genome of *Drechmeria coniospora.* Sci Rep 2016; 6(1): 23122.
[http://dx.doi.org/10.1038/srep23122] [PMID: 26975455]

[43] Khanam S. Characterisation of the interaction between rice and the parasitic nematode *Ditylenchus angustus*, 2016.

[44] Agati G, Biricolti S, Guidi L, Ferrini F, Fini A, Tattini M. The biosynthesis of flavonoids is enhanced similarly by UV radiation and root zone salinity in *L. vulgare* leaves. J Plant Physiol 2011; 168(3): 204-12.
[http://dx.doi.org/10.1016/j.jplph.2010.07.016] [PMID: 20850892]

[45] Bahraminejad S. Biological activity of secondary metabolites in oat (*Avena sativa*), 2007.

[46] Walters D. Plant defense: warding off attack by pathogens, herbivores and parasitic plants. John Wiley & Sons 2011.

[47] Chin S, Behm CA, Mathesius U. Functions of flavonoids in plant–nematode interactions. Plants 2018; 7(4): 85.
[http://dx.doi.org/10.3390/plants7040085] [PMID: 30326617]

[48] Khan F, Pandey E, Fatima S, Khan A, Zeb SZ, Ahmad F. Prospects for the use of metabolomics engineering in exploring and harnessing chemical signaling in root galls. In: Ahmad F, Blázquez GN, Eds, Root-Galling disease of vegetable plants. Springer 2023; pp. 309-38.
[http://dx.doi.org/10.1007/978-981-99-3892-6_13]

[49] Kyndt T, Goverse A, Haegeman A, *et al.* Redirection of auxin flow in *Arabidopsis thaliana* roots after infection by root-knot nematodes. J Exp Bot 2016; 67(15): 4559-70.
[http://dx.doi.org/10.1093/jxb/erw230] [PMID: 27312670]

[50] Fabiyi OA, Baker MT, Olatunji GA. Application of fatty acid esters on meloidogyne incognita infected jew's mallow plants. Pak J Nematol 2022; 40(2) Available from: http://researcherslinks.com/current-issues/Application-Fatty-Acid-Esters-Meloidogyne-incognita-Infected/30/1/5717/html
[http://dx.doi.org/10.17582/journal.pjn/2022/40.2.127.137]

[51] Dahlin P, Hallmann J. New insights on the role of allyl isothiocyanate in controlling the root knot nematode *Meloidogyne hapla.* Plants 2020; 9(5): 603.
[http://dx.doi.org/10.3390/plants9050603] [PMID: 32397380]

[52] Khairy D, Osman MA, Mostafa FAM. Combined use of aqueous plant extracts for controlling *Meloidogyne incognita* and modulating chemical constituents in tomato under greenhouse conditions. Pak J Nematol 2022; 40(1) Available from: http://researcherslinks.com/current-issues/Combined-Aqueous-Plant-Extracts-Controlling-Meloidogyne-incognita/30/1/4857/html
[http://dx.doi.org/10.17582/journal.pjn/2022/40.1.1.11]

[53] Ali S, Naz I, Alamzeb M, URRASHIDd M. Activity guided isolation of nematicidal constituents from the roots of *Berberis brevissima* Jafri and *Berberis parkeriana* Schneid. Tarim Bilim Derg 2019; 25(1): 108-14.
[http://dx.doi.org/10.15832/ankutbd.539012]

[54] Atolani O, Fabiyi OA, Olatunji GA. Nematicidal isochromane glycoside from *Kigelia pinnata* leaves. Acta Agric Slov 2014; 104(1): 25-31. Available from: https://journals.uni-lj.si/aas/article/view/12567
[http://dx.doi.org/10.14720/aas.2014.104.1.3]

[55]   Giri B, Rawat R, Saxena G, Manchanda P, Wu QS, Sharma A. Effect of *Rhizoglomus fasciculatum* and *Paecilomyces lilacinus* in the biocontrol of root-knot nematode, *Meloidogyne incognita* in *Capsicum annuum* L. Commun Integr Biol 2022; 15(1): 75-87.
[http://dx.doi.org/10.1080/19420889.2021.2025195] [PMID: 35273677]

[56]   Silva MF, Paulo Campos V, Barros AF, *et al.* Medicinal plant volatiles applied against the root-knot nematode *Meloidogyne incognita.* Crop Prot 2020; 130: 105057.
[http://dx.doi.org/10.1016/j.cropro.2019.105057]

[57]   Eloh K, Kpegba K, Sasanelli N, Koumaglo HK, Caboni P. Nematicidal activity of some essential plant oils from tropical West Africa. Int J Pest Manage 2020; 66(2): 131-41.
[http://dx.doi.org/10.1080/09670874.2019.1576950]

[58]   Silva JCP, Campos VP, Barros AF, *et al.* Plant volatiles reduce the viability of the root-knot nematode *Meloidogyne incognita* either directly or when retained in water. Plant Dis 2018; 102(11): 2170-9.
[http://dx.doi.org/10.1094/PDIS-01-18-0143-RE] [PMID: 30207900]

[59]   Sandhanam SD, Ganesan P, Stalin A, *et al.* Effect of compound isolated from *Lawsonia inermis* (L.) (Myrtales: Lythraceae) on the immature stages of filarial vector *Culex quinquefasciatus* Say (Diptera: Culicidae) and its docking analysis with Acetylcholinesterase (AChE1). Biocatal Agric Biotechnol 2018; 15: 210-8.
[http://dx.doi.org/10.1016/j.bcab.2018.06.004]

[60]   Ntalli NG, Menkissoglu-Spiroudi U, Giannakou I. Nematicidal activity of powder and extracts of *Melia azedarach* fruits against *Meloidogyne incognita.* Ann Appl Biol 2010; 156(2): 309-17.
[http://dx.doi.org/10.1111/j.1744-7348.2009.00388.x]

[61]   Basaid K, Chebli B, Mayad EH, *et al.* Biological activities of essential oils and lipopeptides applied to control plant pests and diseases: a review. Int J Pest Manage 2021; 67(2): 155-77.
[http://dx.doi.org/10.1080/09670874.2019.1707327]

[62]   Caboni P, Saba M, Tocco G, *et al.* Nematicidal activity of mint aqueous extracts against the root-knot nematode *Meloidogyne incognita.* J Agric Food Chem 2013; 61(41): 9784-8.
[http://dx.doi.org/10.1021/jf403684h] [PMID: 24050256]

[63]   Nguyen LTT, Jang JY, Kim TY, *et al.* Nematicidal activity of verrucarin A and roridin A isolated from *Myrothecium verrucaria* against *Meloidogyne incognita.* Pestic Biochem Physiol 2018; 148: 133-43.
[http://dx.doi.org/10.1016/j.pestbp.2018.04.012] [PMID: 29891364]

[64]   Bothon FTD, Atindéhou MM, Koudoro YA, Lagnika L, Avlessi F. *Parkia biglobosa* fruit husks: Phytochemistry, antibacterial, and free radical scavenging activities. Am J Plant Sci 2023; 14(2): 150-61.
[http://dx.doi.org/10.4236/ajps.2023.142012]

[65]   Laquale S, Candido V, Avato P, Argentieri MP, D'Addabbo T. Essential oils as soil biofumigants for the control of the root-knot nematode *Meloidogyne incognita* on tomato. Ann Appl Biol 2015; 167(2): 217-24.
[http://dx.doi.org/10.1111/aab.12221]

[66]   Lo SC. Physiological and molecular basis of phytoalexin biosynthesis in sorghum. 1999.

[67]   Tholl D. Biosynthesis and biological functions of terpenoids in plants. Biotechnol Isoprenoids 2015; pp. 63-106.
[http://dx.doi.org/10.1007/10_2014_295]

[68]   Ashour M, Wink M, Gershenzon J. Biochemistry of terpenoids: monoterpenes, sesquiterpenes and diterpenes Annu Plant Rev. In: Wink M, Ed., Biochemistry of plant secondary metabolism, Second Edition 2010; 40: pp. 258-303.

[69]   Mrudulakumari Vasudevan U, Lee EY. Flavonoids, terpenoids, and polyketide antibiotics: Role of glycosylation and biocatalytic tactics in engineering glycosylation. Biotechnol Adv 2020; 41: 107550.

[http://dx.doi.org/10.1016/j.biotechadv.2020.107550] [PMID: 32360984]

[70]   Bartnik M, Facey P. Glycosides. In: Badal McCreath SB, Clement YN, Eds. Pharmacognosy. Elsevier 2024; pp. 103-65.
[http://dx.doi.org/10.1016/B978-0-443-18657-8.00001-3]

[71]   Royer M, Houde R, Stevanovic T. Non-wood forest products based on extractives-a new opportunity for Canadian forest industry part 2-softwood forest species. J Food Res 2013; 2(5): 164.
[http://dx.doi.org/10.5539/jfr.v2n5p164]

[72]   Lehari K, Kumar D. Metabolic engineering and production of secondary metabolites. In: Kumar A,Kumar S, Eds, Secondary metabolites and biotherapeutics. Elsevier 2024; pp. 215-44.
[http://dx.doi.org/10.1016/B978-0-443-16158-2.00004-5]

[73]   Pattanaik B, Lindberg P. Terpenoids and their biosynthesis in cyanobacteria. Life (Basel) 2015; 5(1): 269-93.
[http://dx.doi.org/10.3390/life5010269] [PMID: 25615610]

[74]   Baunach M, Franke J, Hertweck C. Terpenoid biosynthesis off the beaten track: unconventional cyclases and their impact on biomimetic synthesis. Angew Chem Int Ed 2015; 54(9): 2604-26.
[http://dx.doi.org/10.1002/anie.201407883] [PMID: 25488271]

[75]   Moore BD, Andrew RL, Külheim C, Foley WJ. Explaining intraspecific diversity in plant secondary metabolites in an ecological context. New Phytol 2014; 201(3): 733-50.
[http://dx.doi.org/10.1111/nph.12526] [PMID: 24117919]

[76]   Caprioglio D, Salamone S, Pollastro F, Minassi A. Biomimetic approaches to the synthesis of natural disesquiterpenoids: An update. Plants 2021; 10(4): 677.
[http://dx.doi.org/10.3390/plants10040677] [PMID: 33916090]

[77]   Kihika RM. Identification of semiochemical mediating root-knot nematode (*Meloidogyne incognita*)-PEPPER (*Capsicum annum*). Interaction 2017.

[78]   Hajianfar R. Analysis of the genetic background of resistance in potato with special attention to late blight (*P. infestans*) resistance= A rezisztencia genetikai hátterének vizsgálata a burgonyában különös tekintettel a burgonyavész (*P. infestans*) rezisztenciára, 2014.

[79]   Winders JR. Discovery of antifungal metabolites in maize cob *via* liquid chromatography-mass spectrometry. Mississippi State University 2019.

[80]   Schmelz EA, Huffaker A, Sims JW, *et al.* Biosynthesis, elicitation and roles of monocot terpenoid phytoalexins. Plant J 2014; 79(4): 659-78.
[http://dx.doi.org/10.1111/tpj.12436] [PMID: 24450747]

[81]   Selim ME, Mahdy ME, Sorial ME, Dababat AA, Sikora RA. Biological and chemical dependent systemic resistance and their significance for the control of root-knot nematodes. Nematology 2014; 16(8): 917-27.
[http://dx.doi.org/10.1163/15685411-00002818]

[82]   Bhatla SC, Lal MA. Secondary metabolites. In: Bhatla SC, Lal MA, Eds, Plant physiology, development and metabolism. Springer 2023; pp. 765-808.
[http://dx.doi.org/10.1007/978-981-99-5736-1_33]

[83]   Abdolabad AS. Characterization of tomato genes for resistance to *Oidium neolycopersici*. Wageningen University and Research 2011.
[http://dx.doi.org/10.18174/164144]

[84]   Zeiss DR. Induced secondary metabolites characterising the defence metabolome in *Solanum lycopersicum* plants responding to *Ralstonia solanacearum* and pathogen-derived elicitors. South Africa: University of Johannesburg 2020.

[85]   Ochola J, Cortada L, Ng'ang'a M, Hassanali A, Coyne D, Torto B. Mediation of potato–potato cyst nematode, *G. rostochiensis* interaction by specific root exudate compounds. Front Plant Sci 2020; 11:

649.
[http://dx.doi.org/10.3389/fpls.2020.00649] [PMID: 32587595]

[86] Wood C. The role of basal resistance in *Arabidopsis thaliana* against the parasitic weed *Striga gesnerioides*, 2020.

[87] Hama JR, Fomsgaard IS, Topalović O, Vestergård M. Root uptake of cereal benzoxazinoids grants resistance to root-knot nematode invasion in white clover. Plant Physiol Biochem 2024; 210: 108636.
[http://dx.doi.org/10.1016/j.plaphy.2024.108636] [PMID: 38657547]

[88] Ngala BM. The use of brassica species for the management of potato cyst nematode infestations of potatoes, 2015.

[89] Potter MJ, Vanstone VA, Davies KA, Rathjen AJ. Breeding to increase the concentration of 2-phenylethyl glucosinolate in the roots of *Brassica napus*. J Chem Ecol. 2000 Aug; 26: 1811-20.
[http://dx.doi.org/10.1023/A:1005588405774]

[90] Musa N. Management of the stem and bulb nematode (*Ditylenchus* spp.) in winter beans (*Vicia faba* L) using biofumigant *Brassica* spp. and other allelopathic cover crops, 2022.

[91] Gommers FJ, Voorin'tholt DJM. Chemotaxonomy of Compositae related to their host suitability for *Pratylenchus penetrans*. Neth J Plant Pathol 1976; 82(1): 1-8.
[http://dx.doi.org/10.1007/BF01977341]

[92] Arroo RR. Regulation of thiophene biosynthesis in *Tagetes patula* L. [Sl: sn]; 1994.

[93] Takasugi M, Yachida Y, Anetai M, Masamune T, Kegasawa K. Identification of asparagusic acid as a nematicide occurring naturally in the roots of asparagus. Chem Lett 1975; 4(1): 43-4.
[http://dx.doi.org/10.1246/cl.1975.43]

[94] Parvaiz S, Yousuf P, Lone R, Rather YA. Plant phenolics in alleviating root-knot disease in plants caused by *Meloidogyne* spp. In: Lone R, Khan S, Al-Sadi AM, Eds, Plant Phenolics in Biotic Stress Management. Springer 2024; pp. 417-39.
[http://dx.doi.org/10.1007/978-981-99-3334-1_17]

[95] Razzaq A, Sadia B, Raza A, Khalid Hameed M, Saleem F. Metabolomics: A way forward for crop improvement. Metabolites 2019; 9(12): 303.
[http://dx.doi.org/10.3390/metabo9120303] [PMID: 31847393]

[96] Gillet FX, Bournaud C, Antonino de Souza Júnior JD, Grossi-de-Sa MF. Plant-parasitic nematodes: towards understanding molecular players in stress responses. Annals of Botany. 2017 Mar 1;119(5):775-89.
[http://dx.doi.org/10.1093/aob/mcw260] [PMID: 28087659]

[97] Sikder MM, Vestergård M, Kyndt T, Topalović O, Kudjordjie EN, Nicolaisen M. Genetic disruption of *Arabidopsis* secondary metabolite synthesis leads to microbiome-mediated modulation of nematode invasion. ISME J 2022; 16(9): 2230-41.
[http://dx.doi.org/10.1038/s41396-022-01276-x] [PMID: 35760884]

[98] Serag A, Salem MA, Gong S, Wu JL, Farag MA. Decoding metabolic reprogramming in plants under pathogen attacks, a comprehensive review of emerging metabolomics technologies to maximize their applications. Metabolites 2023; 13(3): 424.
[http://dx.doi.org/10.3390/metabo13030424] [PMID: 36984864]

[99] Huang B, Wang J, Han X, *et al.* The relationship between material transformation, microbial community and amino acids and alkaloid metabolites in the mushroom residue-prickly ash seed oil meal composting with biocontrol agent addition. Bioresour Technol 2022; 350: 126913.
[http://dx.doi.org/10.1016/j.biortech.2022.126913] [PMID: 35231600]

[100] Gathungu RM, Kautz R, Kristal BS, Bird SS, Vouros P. The integration of LC-MS and NMR for the analysis of low molecular weight trace analytes in complex matrices. Mass Spectrom Rev 2020; 39(1-2): 35-54.
[http://dx.doi.org/10.1002/mas.21575] [PMID: 30024655]

[101] Nguyen VPT, Stewart J, Lopez M, Ioannou I, Allais F. Glucosinolates: natural occurrence, biosynthesis, accessibility, isolation, structures, and biological activities. Molecules 2020; 25(19): 4537.
[http://dx.doi.org/10.3390/molecules25194537] [PMID: 33022970]

[102] Nyczepir AP, Thomas SH. 18 current and future management strategies in intensive crop production systems. Root-Knot Nematodes 2009; p. 412.

[103] Abd-Elgawad M, Molinari S. Markers of plant resistance to nematodes: classical and molecular strategies. Nematol Mediterr 2008.

[104] Pieterse CMJ, Zamioudis C, Berendsen RL, Weller DM, Van Wees SCM, Bakker PAHM. Induced systemic resistance by beneficial microbes. Annu Rev Phytopathol 2014; 52(1): 347-75.
[http://dx.doi.org/10.1146/annurev-phyto-082712-102340] [PMID: 24906124]

[105] Molinari S, Baser N. Induction of resistance to root-knot nematodes by SAR elicitors in tomato. Crop Protection. 2010 Nov 1;29(11):1354-62
[http://dx.doi.org/10.1016/j.cropro.2010.07.012]

[106] Choudhary P, Aggarwal PR, Rana S, Nagarathnam R, Muthamilarasan M. Molecular and metabolomic interventions for identifying potential bioactive molecules to mitigate diseases and their impacts on crop plants. Physiol Mol Plant Pathol 2021; 114: 101624.
[http://dx.doi.org/10.1016/j.pmpp.2021.101624]

[107] Hajjar DA. Bio-prospecting of plants and marine organisms in saudi arabia for new potential bioactivity [Internet]. KAUST Research Repository; 2016. Available from: https://repository.kaust.edu.sa/handle/10754/621995

[108] Rush TA, Shrestha HK, Gopalakrishnan Meena M, *et al.* Bioprospecting *Trichoderma*: A systematic roadmap to screen genomes and natural products for biocontrol applications. Front Fungal Biol 2021; 2: 716511.
[http://dx.doi.org/10.3389/ffunb.2021.716511] [PMID: 37744103]

SUBJECT INDEX

Z

Zinc finger 15